Project Control

Integrating Cost and Schedule in Construction

Second Edition

Wayne J. Del Pico
WJ Del Pico Inc.
Pembroke, MA, USA

Published by John Wiley & Sons, Inc., Hoboken, New Jersey.
Published simultaneously in Canada.

Library of Congress Cataloging-in-Publication Data:

Names: Del Pico, Wayne J., author. | John Wiley & Sons, publisher.
Title: Project control : integrating cost and schedule in construction / Wayne J. Del Pico.
Description: Second edition. | Hoboken, New Jersey : Wiley, [2023] | Includes index.
Identifiers: LCCN 2023007175 (print) | LCCN 2023007176 (ebook) | ISBN 9781394150120 (paperback) | ISBN 9781394150144 (adobe pdf) | ISBN 9781394150137 (epub)
Subjects: LCSH: Construction projects—Management. | Building—Estimates. | Scheduling.
Classification: LCC TH438 .D448 2023 (print) | LCC TH438 (ebook) | DDC 624.068/4—dc23/eng/20230302
LC record available at https://lccn.loc.gov/2023007175
LC ebook record available at https://lccn.loc.gov/2023007176

Cover Design: Wiley
Cover Image: © Jung Getty/Getty Images

Set in 11/14pt Times LT Std by Straive, Chennai, India

SKY10051919_072623

Project Control

Arman J. Del Pico
(1921–2011)

To my father, who taught me by quiet example the meaning of living a life based on a personal code of values and strength of character.

Contents

About the Author

Wayne J. Del Pico is President of W.J. Del Pico, Inc., where he provides consulting and litigation support services for construction-related matters. He has more than 43 years of experience in construction project management and estimating and has been involved in projects throughout most of the United States. His professional experience includes private commercial construction, public construction, retail construction, and residential land development and construction.

Mr. Del Pico holds a degree in civil engineering from Northeastern University in Boston, where he taught construction-related curriculums in Cost Estimating, Project Management, and Project Scheduling from 1992 until 2006. He is also a member of the adjunct faculty at Wentworth Institute of Technology in Boston, where he has taught programs in Construction Cost Analysis, Estimating, Project Control, and Construction Scheduling since 2003.

Mr. Del Pico is a seminar presenter for Gordian where he lectures on estimating topics using RSMeans data. He holds the designation of RSMeans Data Certified Professional.

Mr. Del Pico is the author of *Plan Reading and Material Takeoff* (1994) and *Estimating Building Costs* (2004, 2012, and 2023), *Project Control: Integrating Cost and Schedule in Construction* (2013 and 2023), and *Electrical Estimating Methods* (2015), and is a coauthor of *The Practice of Cost Segregation Analysis* (2005).

He has served as the 2010 president of the Builders Association of Greater Boston. He is also a practicing Neutral for the American Arbitration Association since 2009, where he hears construction-related arbitration cases.

Preface

When the first edition of *Project Control: Integrating Cost and Schedule in Construction* was written in 2013, it was out of a need for a specific text on just project control. At that time I had been teaching a project control class at Wentworth Institute of Technology for about 5 or 6 years and had used a few different texts supplemented by a series of papers to fill in the gaps. While they were all excellent texts by accredited authors, none of them dealt with the topic of project control exclusively. Most had a few chapters in a predominantly scheduling or project management text. The few texts that did, were not construction-specific, but applicable to any type of project control from software projects to manufacturing. They were very general. I felt it was imperative to have a text that just focused on our industry. The aim was to make it more relatable for construction professionals.

The second thing I noticed with these other texts was the writing style. They were all written with a very "starched," polished style typical of academia. Some were just difficult to read, especially for students. My objective was to make *Project Control* more user-friendly and easier to read and understand. Let's face it, while the topic is interesting if you are in the profession, it's not exactly a page turner. *Project Control* was intentionally written in a lighter style without sacrificing content or clarity.

Over the course of the last ten years, *Project Control: Integrating Cost and Schedule in Construction* has met with tremendous, but unexpected success. It has been adopted by numerous trade schools, colleges, and universities throughout the United States as a mainstay in their construction curriculum. While its biggest appeal has been as a teaching text, it has also found popularity as a desk reference for the novice or experienced construction professional looking to improve their project controls. As a result, the publisher has requested a second edition.

This second edition has maintained the same writing style, but has been expanded to incorporate comments from both teachers and students as to what they felt could be more thoroughly explained. It also includes a revised section of questions and answers for each chapter. New to the second edition is a case study with a series of questions that can be used for group discussion. The second edition has also included changes due to time.

I sincerely hope that you will enjoy this second edition and your career will benefit from reading it.

—Wayne J. Del Pico

January 2023

CHAPTER 1

THE BASICS

Even for the most seasoned professional with years of construction project management experience, it is still beneficial to revisit the basics once in a while, if for no other reason than to "reinforce" what we already know. This chapter provides an overview of project management theory and shows where project control fits into the discipline. It does not offer detailed instruction in the profession of day-to-day project management, but will review the theory of project management and the basic concepts common to the construction industry. The focus will be from the viewpoint of the project manager of the contractor.

The Concept of Project Management

Project management is the professional discipline of planning, monitoring, and controlling specific resources to achieve a predetermined set of goals for a project. The term *project* is defined as a one-time endeavor with well-defined, often unique, goals, with specific limits of both time and cost to achieve those goals. The temporary nature of a project is in contrast to businesses like manufacturing that are repetitive by nature. This type of management has predefined systems or processes that have been refined over time. The management of a project, in contrast to manufacturing, requires a different paradigm and skill set on the part of the practitioner. The project manager's main test is to achieve all of the goals, including the deliverables, within the constraint of a fixed budget and fixed time frame. Specific goals and deliverables in the construction industry will vary from project to project, but will always be within the parameters of satisfying the contractual objectives of time and cost.

For the purpose of this text, let's refine the definition to reflect a project as defined by the construction industry. Project management in the construction industry is the professional practice of planning, scheduling, monitoring, and controlling a finite amount of resources—materials, labor, equipment, and subcontractors—to achieve a set of goals, which are usually a series of engineering improvements for a unique one-time event, the

Project Control: Integrating Cost and Schedule in Construction, Second Edition. Wayne J. Del Pico.

project. To compound this definition, planning, scheduling, monitoring, and controlling must occur within a defined duration and a frequently fixed budget. To further complicate the process, let's add some unpredictable factors such as weather, labor strikes, material shortages, supply chain variables, and a wide range of separate goals and conflicting agendas proposed by the participants. Lastly, let's not forget the ever-present threat of financial penalty as a "reward" for failing to master all of the above challenges!

History of Project Management

The concept of project management is not new. It has been practiced since human beings noticed the need for improvement in their surroundings. There is a near 100 percent chance that the construction of the Great Pyramids of Giza had a project manager or multiple project managers over time. Rather than bore the reader with recounting the development of project management since the dawn of the Bronze Age, let's consider what's important.

As a professional discipline, project management emerged as construction projects became more complex and warranted an individual to be accountable for the performance and results. Techniques for managing projects started to develop at the dawn of the twentieth century.

Much of the credit for the development of project management techniques can be given to two early pioneers: Henry L. Gantt and Henri Fayol. Both Gantt and Fayol were students of Frederick Winslow Taylor's theories of scientific management.

Henry L. Gantt (1861–1919) was an American mechanical engineer who later became a management consultant to the steel industry. He worked with Taylor until 1893, applying scientific management principles to the production of steel. Gantt is called the father of planning and control techniques in project management. He is most recognized for the development of a visual management tool that displays a task as a function of time. It is used to measure actual progress against planned progress. The tool was aptly named the *Gantt Chart* and is still very much in use today.

Gantt's 1919 book, *Organizing for Work*, describes two principles of the Gantt Chart:

- Measuring activities by the amount of time needed to complete them (task duration)
- Representing the quantity of the task that should have been done in that time (daily output)

Both of these principles have a direct applicability to construction. Gantt was one of the first to recognize that in order for a team of workers to produce efficiently and to maintainable standards, the team needed an intelligent leader who could solve or preempt problems, and provide guidance to control the outcomes of certain actions.

Henri Fayol (1841–1925) was a French mining engineer who developed a general theory of business administration later referred to as *Fayolism.* Fayol believed in analyzing the role of management to reduce problems and to increase worker efficiency. Fayol is credited with the creation of the six management functions that are still the basis of project management today;

- Forecasting
- Planning
- Organizing
- Commanding
- Coordinating
- Monitoring

Fayol's book, *General and Industrial Management*, published in 1916, outlined a flexible theory of management that could be applied to most industries. The book stressed the importance of planning as a way to improve efficiency and control the outcomes.

With the dawn of the 1950s and the increase of the military-industrial complex to offset the threat of the Cold War, modern project management really came to life. Project management became a recognized stand-alone professional discipline. This decade saw the development of two mathematical models for project scheduling that added to the project management toolbox. The *Critical Path Method*, or CPM as it is more commonly known, was developed for managing maintenance projects by two U.S. industrial giants; DuPont and Remington Rand. The second method, *Program Evaluation and Review Technique* (PERT), was developed as part of the U.S. Navy's Polaris missile submarine program. These mathematical techniques for scheduling were quickly adopted by the construction industry.

At the same time these scheduling models were being developed, advances in technology were evolving in cost estimating and cost management and control. Combined, these techniques gave birth to the discipline of project control.

Approaches to Project Management

There are several approaches to managing project activities, each with special characteristics. Depending on the particular perspective of the project manager (owner, design professional, constructor, etc.), one approach may be better suited than another. The approaches are as follows:

- **Traditional Approach:** The Traditional Approach identifies a series of phases that drive the project. In the Traditional Approach there are five distinct steps:
  - Initiation
  - Planning
  - Execution
  - Monitoring and Controlling
  - Completion

The Traditional Approach and the individual processes within it will be discussed in more detail later in this text.

- **Critical Chain Project Management (CCPM):** CCPM is a fairly new technique of project management that puts more emphasis on the resources needed to do the project tasks. It is an application of the *Theory of Constraints* (TOC) where the goal is to increase the completion rates of projects within an organization. There are five steps that CCPM focuses on:
  - Identify the constraint.
  - Exploit the constraint (do what is required to ensure the constraint works at optimum capacity).
  - Subordinate all other tasks to the constraint (the constraint is given priority).
  - Elevate the constraint (obtain more of the constraint).
  - Repeat the cycle.
- **Extreme Project Management (XPM):** XPM is reserved for very complex and uncertain projects often associated with software development. It focuses on the management of human resources versus formal scheduling techniques.
- **Agile Project Management (APM):** APM employs a nontraditional methodology similar to XPM. It requires a complete rethinking of the traditional processes. APM has its greatest applicability in the software development business, but has also been adapted to complex construction and design projects.
- **Event Chain Methodology:** This is an uncertainty modeling and schedule analysis technique that focuses on the management of events and event chains that affect the project schedule. It is often a complement to the CPM schedule technique. Event chain methodology is used to perform more accurate quantitative analysis while considering such factors as relationships between different events and actual timing of the events.

As one can imagine, not all of the aforementioned methodologies are applicable to the construction industry. Regardless of the approach selected, a detailed analysis of the overall goals of the project, the available time for performance, the budget, and the roles and responsibilities of the participants is primary.

The Traditional Approach

As an overall methodology for the average construction project, the Traditional Approach has the greatest applicability. This methodology is often called the waterfall approach because of its linear execution: the predecessor task must be completed before starting the successor task.

If one were to analyze each of the five steps in the process, it would quickly become apparent where the steps fit in a construction project.

The Initiating Process

The *Initiating Process* determines the nature and scope of the project. What is the need, will this project satisfy that need, and can the need be accommodated within the client's budget? This is where the client's expectation of what they will receive, in the form of the project, will be tempered by what the client can afford. This particular phase must be done correctly as the rest of the program will be driven by its outcomes. Incorrect assumptions about the business model, industry growth, or the economic environment, to name but a few, may result in a project that does not satisfy the client's need.

The Initiating Process should address the following:

- Define the current business model.
- Review current practices or operations.
- Define expectations for growth.
- Analyze the business needs in measurable terms.
- Obtain consensus on needs from stakeholders/end users.
- Analyze financial cost and benefits.
- Decide to go forward with the project.

It is common for the design team or the construction manager to perform the majority of the processes in the initiating phase with the owner's concurrence. This may be part of the feasibility analysis of the program long before the project goes out for bid to contractors.

The Planning Process

The *Planning Process*, sometimes called the planning and design process, investigates and evaluates the best method in which to achieve the expectations (goals) defined in the Initiating Process. The Planning Process further identifies the scope and parameters of the project: time, cost, and resources. The Planning Process often identifies the team and its hierarchy.

The Planning Process includes the following steps:

- Select the planning team (in-house and/or outsourced).
- Assign responsibilities to team members and contributors.
- Develop the scope of work statement.
- Identify the resources required to perform steps in the plan and the availability of those resources.
- Identify and evaluate potential means and methods for the project.
- Estimate the cost to achieve the goals: the budget.
- Identify the time available for performance.
- Break the deliverables down into phases (if appropriate).

- Perform a risk analysis and create a risk management plan as a result of the analysis.
- Decide to go forward with the project.

The success of the Planning Process often defines the success of the project as a whole. A separate way to define the Planning Process at the contractor level would be to consider the brainstorming and planning that go on between project manager team members when they create the plan and determine how the work will be executed. The goal is to arrive at the *means and methods* for performing the work. Means and methods is the "game plan" for performing the work, and this is the expertise that the contractor brings to the table. While the contract is required to be executed in conformance with the plans and specifications, there are different ways in which the work can be performed. The goal of the plan is to execute the work in the most cost-efficient and timely manner. It is the balance between time and cost.

The Executing Process

The *Executing Process* is where the Planning Process is put into action. It is where the work gets done and resources strive to achieve the project goals. Executing involves coordination of the resources to meet the project deliverables within the time and budget. It is a combination of leadership and management techniques aimed at getting results. It is also where the individuals managing the process advise the management and planning team of what portion of the plan works and what portion of it needs to be adjusted. This is called the *feedback cycle* and is crucial to achieving the project goals. The feedback cycle will be discussed in more detail later in this text.

The Executing Process should include the followings steps:

- Select or contract with the resources that will perform the work.
- Select a methodology for the individual critical tasks.
- Develop a schedule (CPM) for performance of the work.
- Execute the tasks in accordance with the schedule.
- Execute the tasks in accordance with the plan.
- Establish the metrics for performance measurements (create a baseline to measure performance against).

While an argument could be made for several of the previous steps to occur in the Planning Process, the team that does the planning may be different than the team that performs the work. Therefore, as professionals, those who perform the work should have some latitude as to what means or method works best. The plan is often refined during the Executing Process. However, it is not revised at the cost of losing sight of the efficiency goals.

The Monitoring and Controlling Process

The Monitoring and Controlling Processes, while separate, are performed together. They consist of a series of steps to observe the executing process. It is the establishment of a guidance system or a set of metrics by which actual performance can be measured and compared to planned performance. It is the part of the project where the differences between planned and actual performance in both cost and schedule, called *variances*, are analyzed. If required, a series of inputs called *corrective actions* are implemented to achieve the original desired outcome based on the plan. They help to guide the project back on to the track of the original plan. This is a crucial step in the process, without which it is impossible to determine how well the project is performing.

The essence of any type of management is control. It is fundamental to project management; if one is to manage, one must control. *Wideman Comparative Glossary of Project Management Terms* v3.1 defines control as: "The process of comparing actual performance with planned performance, analyzing the differences, and taking appropriate corrective action." This process, if performed correctly, offers the distinct benefit of knowing the project status at all times.

The Monitoring and Controlling Process includes:

- Updating metrics for performance measurements (baseline).
- Measuring the performance of ongoing tasks (during the Executing Process).
- Monitoring of the project variables (cost, time, resources, and quality) against the baseline.
- Analyzing the variances between baseline and actual, and their cause.
- Identifying and implementing corrective actions to get the actual back on track.

It also includes a variety of ancillary tasks that become part of the process, such as change management and value engineering. The performance baseline identified in the first step is often developed in the Executing Process and updated or adjusted as the work progresses.

The Closing Process

The process for formal acceptance at or near the end of the project is called the *Closing Process*. In the construction industry, this is called "Project Closeout." In the framework of the construction execution, it is ongoing from Substantial Completion (beneficial use and occupancy) to Final Completion.

The Closing Process includes:

- Development and completion of punchlist tasks.
- Finalizing of record documents.

- Training of owner personnel.
- Turnover of operations and maintenance manuals.
- Archiving of project files.
- Accounting and closeout of a contract.
- Conducting a lessons-learned meeting.

This is a critical part of the overall project and a necessary step to a successful project.

If one were to follow the aforementioned processes through a construction project, it is easy to imagine how the client (owner) would use each process to expand its operations by constructing a new manufacturing facility. It could also be applied to the general contractor recently awarded the contract to construct the facility or to the design firm awarded the contract for the documents. Each member of the team performs similar steps to manage their portion of the project.

Benefits of Proper Planning

Each process in the Traditional Approach to project management is crucial to its overall success. Each step has distinct contribution to the next step in the process that allows that next step to build upon the previous one. This is referred to as the *waterfall model*. The waterfall model is successful in the rigid environment of the construction industry where changes after the fact can be cost-prohibitive. However, for the construction professional, each step or process may hold a different weight based on its contribution.

For example, the architects and engineers charged with designing the project place a heavier weight on understanding the client's needs and translating that into a program that can be conveyed to those who must construct it. Like most professions, architects and engineers can specialize in the design of hospitals, offices, churches, and just about any unique use a structure could have. It is a result of knowing the proper questions to ask the client and how the responses can be put on paper. However, if they do not understand the client's needs, the design is destined for failure.

Contractors that will construct the project may place more weight on the Planning Process. Proper planning forces detailed thinking about the project. It allows the project manager (or team) to "build the project in his or her head." The project manager (or team) can consider different methodologies, thereby deciding what works best or what does not work at all. This detailed thinking may be the only way to discover constraints or risks that were not addressed in the estimating process. It would be far better to discover in the planning phase that a particular technology or material will not work than in the Execution Process.

The goal of the Planning Process for the contractor is to produce a workable scheme that uses the resources efficiently within the allowable time and allotted budget. A well-developed plan does not guarantee that the Executing Process will proceed flawlessly or that the project will even succeed in meeting its objectives. It does, however, greatly improve its chances! The most significant contribution of the Planning Process is the baseline that is essential in the performance of project control. Without a thoroughly developed plan, what would the contractor use to measure progress?

When the plan is set to time, it becomes the schedule. Construction scheduling is done by professionals who are familiar with the construction process as it is performed in the field. It requires an in-depth knowledge of construction methods and an ability to visualize tasks and their interdependencies with other tasks. Knowledge is gained in the Planning Process that is beneficial in developing the schedule.

While the schedule does not require a consensus of all of the participants, it does require input from those who will perform or manage the work in the field. Superintendents, foremen, subcontractors, and vendors can add valuable insight and perspective as to how the work should be performed. Subcontractors can often flush out the details to fit within the prime contractor's schedule. Allowing the participants to contribute not only adds to the knowledge base about the project, but it also creates *buy-in*. Buy-in is the agreement by a project participant to support an idea or decision because they were allowed to contribute to the formation of that idea or decision. Buy-in is an essential element to a project's success. When a participant offers a suggestion that is then incorporated in the plan or schedule, there is a natural desire by that participant to see his/her suggestion succeed. This desire is often followed up with a series of actions that ensure the success of the suggestion, all of which is the result of proper planning.

The Role of the Project Manager

Up to this point, the book has discussed the concept of project management from a generic point of view. Going forward, it will be helpful to have a particular perspective. The viewpoint will be that of the project manager for the contractor. The author has specifically omitted "sub-," "prime," or "general" before the word "contractor" (except where absolutely necessary) so as not to alienate other frames of reference. With very little imagination, the ideas and concepts presented here are synonymous with most contracting companies in the construction industry.

To understand the role the project manager plays in the contractor's office, it is essential to understand the term *project* as it relates to the construction industry. Construction projects for the most part are unique, dynamic, and complex. In some less than temperate climates, the exact same building constructed on the same property in two different seasons of the year can be totally different projects. Projects are affected by their owners, design professionals, and contractors as well as the relationships that are fostered between the parties. Professionals who cooperate and execute their responsibilities in a timely manner further the interests of the project, and those who do not—hinder its progress. Projects change constantly; they are dynamic in nature. Expecting otherwise is naïve. Managing that change is the essence of successful project management.

Summarizing the previous paragraphs, the role of the project manager in construction can be distilled into three actions: *plan*, *monitor*, and *control*. While there is nothing simple about the actions and skills required to complete each, almost every action can fit into one of those actions.

So how does the project manager fit into the process?

Lead the Project Team

The project manager, or PM, is the point person responsible for the project. He or she is accountable for wins and losses on both the schedule and budget. They are tasked with utilizing the company's money, equipment, human resources, and internal core competencies efficiently. The project manager heads the contractor's project team and is the lead decision-maker. The project team can be defined as the contractor's employees: the supervisor, and tradespersons directly employed by the contractor, subcontractors who will perform work on-site, vendors who will provide specific equipment or fabricated items to the site, and the suppliers who will provide essential components such as ready-mix concrete, lumber, and building materials.

The project manager will lead the team in other ways as well. They include representing the contractor's team in meetings with the owner and architect/engineers, lenders or investors in the project, organized labor groups, and local governmental agencies or committees having jurisdiction over the project. The project manager will be responsible for reporting to account executives or senior management on the company's performance on that project. The PM's decisions and communications will be viewed as the official decisions and communications of the contractor—the legal voice of the contractor.

Create the Project Plan

The project manager leads the team that is responsible for creating the project plan. On some small projects the PM may be the entire team. As previously discussed, the project plan is the workable scheme to accomplish the project's intended goals. The PM must devise the scheme within the confines of the contract documents, contractual constraints, and the available budget as defined by the estimate. It has to be above all a realistic plan that can be achieved by normal humans. The plan has to consider all of the resources that will be required and the most expeditious and efficient use of those resources. Unpredictable events, such as material shortages, supply chain delays, labor strikes, and price escalations, and even the weather must be taken into account in the plan. The plan must have measurable events or deliverables that can be used to assess whether the plan is working, and the team is achieving its goal.

The project manager will assemble, oversee, and assume responsibility for those individuals and companies who are best suited to execute this plan. He or she will be responsible for analyzing which key trades are critical to the success of the project. The PM must identify the areas of high risk and work aggressively to mitigate the risk. Lastly, the project has to meet the parameters set forth in the contract for performance and quality. Above all, the plan has to be accomplished within a definitive timeframe.

Develop the Project Schedule

With the exception of large contracting companies that can afford the luxury of a dedicated project scheduler, the duty of developing a schedule is typically the responsibility of the project manager. Even for the simplest project, the schedule is a complex, ever-changing, and project-specific management tool. The schedule decomposes the project into the individual tasks that comprise the whole. It organizes these tasks in chronological order, and clearly illustrates the interdependencies between them. The schedule enumerates definitive, recognizable milestones or intermediate goals, all of which are set as a function of time.

Each of the tasks on the schedule, referred to as *activities*, has a party responsible to execute it, called the *resource*. Those parties may be subcontractors who have a continued interface on the project, such as the electrician or HVAC contractor, or single application subcontractors who are on and off the project relatively quickly, such as the damp-proofing expert or the toilet partition installer.

In addition to the actual production activities, there are also administrative and procurement components such as submittals and lead times that are incorporated in the schedule and have a direct impact on progress. The schedule should also include activities by outside parties such as the owner or building inspectors whose performance impacts the project's progress. Once the schedule has been accepted by the team, it becomes the baseline for comparison of the actual performance of the work. This is key to project control. It is impossible to determine if a project is ahead or behind if there is nothing to compare it to!

Monitor the Progress of the Project

In order to determine if the project is on schedule and in accordance with the plan, the project manager is required to monitor the progress of the work. This requires more than a cursory review of progress. As the work progresses, the schedule has to be updated to reflect actual performance. The monitoring portion of the Monitoring and Control Process is meant to establish a comparison between the actual progress and the anticipated progress of the project as defined in the schedule. The monitoring reveals which activities are ahead of schedule, on schedule, and, most importantly, behind schedule. The monitoring is based on *feedback*, an essential part of the decision-making process. Feedback comes in many forms: verbal communications, written daily reports from the field, labor-tracking reports, material deliveries, milestones achieved, and upcoming activities. Without the feedback, there can be no effective monitoring or informed decision-making. The monitoring also includes oversight of the costs that are incurred as the work is performed.

Control the Project

One of the most important responsibilities of the project manager is to control the project. Control in most cases can be defined as making decisions proactively instead of reactively to guide the direction of the project. Decisions must be made by considering all of the information available at the time and acting in the best interest of the project. Controlling the project includes adjusting both the plan and the schedule when things change, as they inevitably do. The PM must remember that there are always alternate methods for achieving the project goals. It is impossible to be proactive in every situation; occasionally, crisis management is the topic of the day. Great plans sometime change, and sometimes they just plain fail altogether. Unfortunately, that is just the nature of the business. How quickly and cost-effectively the PM can get the project back on track is the essence of great project management!

Take Corrective Action

When the progress of the project falls short of the plan or deviates from the schedule, the project manager must be willing and able to take immediate and effective action to correct the deviation. The corrective action runs the full spectrum from written directives to termination. The actions necessary are determined as a result of analyzing all of the data in the feedback cycle, discussing the options with team members, and considering any potential consequences of those actions.

It is essential that corrective action is applied in a timely and professional manner. Implementation of a corrective action has one goal, and that is to bring the project back on schedule. It is never intended to be punitive.

Not every action will have the desired or intended effect. In fact, it is not unusual for the corrective action to have no positive impact at all. Recognizing the mistake and switching to Plan B is often the next step in the process. This project manager should not chalk this up as a failure, but as a step in the process of getting the project back on track. Inaction or waiting for the problem to correct itself are often far worse than taking the wrong action.

Achieve the Project Goals

In summary, the main duty of the project manager is to achieve the project goals. The planning, scheduling, monitoring, and controlling of the work are all designed and intended to further the project goals.

Goals of the Project

The goals of the project are the goals of the project team. As the lead on the project team, the goals of the project manager are synergized with the team. They are the result of a clear understanding of the project documents made evident during the Planning Process. Some of the goals are contractual in nature, and some are imposed by the senior management. Understanding those goals and keeping them on target and in the focus of the project team is paramount to the success of any project and a primary responsibility of the project manager. Many projects become hopelessly derailed because the team loses focus of the goals or, worse, were unclear about what the goals were in the beginning.

Contractual Performance Obligations

At the top of the list of the project goals is achieving the contract obligations. Every contract for construction has specific requirements for performance. These include standards for quality of materials and workmanship, specific deliverables, such as phases, and the most significant performance requirement: Substantial Completion. In their simplest terms, contracts define quality, time, and price. It is the primary goal of the project team to meet the performance requirements of the contract. Failure to meet these obligations can render the remaining goals on this list unattainable. The project manager and the project team must have a thorough understanding of the contract requirements and deliverables. There is no shortcut to this step.

Financial Objectives

While it may sound shallow or ignoble of purpose, all project managers are tasked with meeting (or exceeding) the financial goals outlined in the estimate or by the project team. That means the project should contribute its share to the company's financial bottom line. As with any for-profit business enterprise, making a return for managed risk is one of the determinants of project success. Project managers are often the gatekeepers to the profits. They are charged with the task of maintaining the estimated profit carried in the bid, in addition to the profit that comes from negotiating, effective decision-making, controlling the participants, and superior management of the schedule. In the eyes of most construction executives, one of the most important measures of a successful project is the net profit it returns to the company's balance sheet.

Prevent or Minimize Delay

The scariest word in the construction lexicon is "delay." The principal reason is that recovery from delay can be very costly, if even possible. Every project has tasks that fall behind, but with proper guidance are brought back on track. Delay can become chronic when the entire project is behind on the project schedule. Once a project has experienced a delay, it can provoke a host of claims from the affected parties: claims of losses from the owner, claims of material escalation costs, acceleration costs from subs, imposition of liquidated damages, and even additional charges from the architect and engineers for extended services. Resolving delay disputes is time-consuming and distracts the project manager from his or her more immediate responsibilities in managing the project. Careful monitoring of a project schedule and control of the contractors performing the work can prevent or minimize delay. Gaining minor leads and exploiting them can often provide a little cushion to any potential delay.

Avoid Claims or Litigation

Disputes arise on every project. It is in the very essence of what we do. Disputes give rise to claims. Claims for additional work as a result of delay, acceleration or escalation costs are normal occurrences on a construction project and should be dealt with in the life cycle of the project. Those claims that can't be substantiated and approved in the normal course of business often end up with a third party overseeing the resolution of the dispute. Preparing to litigate, arbitrate, or mediate a dispute is a burden on the project manager's time and can be all-consuming. Since most actions to resolve disputes occur well after the incident, the project manager must review and study the project records and become re-acquainted with the details. Avoiding claims or possibly the resulting legal actions is a primary responsibility of the project manager. It can very quickly turn a financially successful project into a loser.

Control the O-P-C Relationship

The O-P-C relationship is the tri-party arrangement most often used in the construction industry under the design/bid/build delivery methodology. The Owner (O) contracts with the Professional (P), most often an architect or engineer, for design and administration services. The Owner (O) also contracts separately with the Contractor (C) to construct the physical project. While there is no direct binding agreement between the professional and the contractor, there is a requirement that they cooperate in the best interest of the owner to complete the project. As the party with the most financial risk, it is the project manager for the general contractor who needs to control the O-P-C relationship as well as the project team. The project manager must strive to earn the respect and trust of the architect and the owner so that he or she can be the proactive force in managing the project, instead of reacting to the wishes of the professional and the owner. A strong PM with a clear understanding of the project and its unique features, who can direct the work and maintain the project schedule, is the hope of every owner and design professional for their project. A PM who can demonstrate these skills and lead instead of follow will often be given a greater latitude by the owner and design professional.

Increase Market Share

In all businesses, a successful engagement, a well-executed contract, or the sale of a product that performs as advertised should increase business and add to a company's market share. The construction industry is no exception. There is only so much advertising that a company can do. Successful performance adds to a company's reputation and ultimately its bottom line. As the old adage goes: "The proof is in the pudding." While it may not be the primary goal of the project manager to increase business, it should be an ancillary goal. It is the by-product of a project done well, and a reward for superior performance by the project manager. It also has the distinct advantage of preventing the PM from "being made available to the industry."

Role of the Contract Documents

The contract documents, sometimes referred to as the CDs, are the plans and specifications, plus any addendum provided during the bidding process. These documents are the basis of the contract. The plans are the quantitative portion of the project, represented graphically. Quantities associated with any activity or material incorporated in the project are determined from the plans. The technical specifications are typically separate from the plans and are bound in the project manual. They define the qualitative portion of the project and address the characteristics of the product and the acceptable tolerances of the workmanship. The plans and specifications are meant to be used together; neither is meant to exist as a stand-alone document. The CDs establish the standard by which the performance of the contractor is measured. Through the plans and specifications, the documents define the quality and parameters for performance. If the contractor has performed in accordance with the contract documents, then the project should be successful from a performance standpoint. The CDs are the most valuable tool that the project manager has in his/her toolbox when managing other team members.

Study the Plans and Specifications

Long before the commencement of any physical work on the site, and prior to subcontracts being let and materials purchased, the project manager should *study* the plans and specifications carefully and thoroughly. As the word implies, study is far more than a casual review. A careful and thorough study of the contract documents is the project manager's education on the project. This is most often done in a quiet setting without interruptions from the phone, e-mail, coworkers, or the normal day-to-day disturbances that demand the project manager's attention. It is helpful to have a pad of paper and pencil for notes as the study takes place.

The project manager should review in detail each and every page of the plans and technical specifications. There are many reasons for the in-depth study, the least of which is to find the design professional's mistakes. While it is implied that the plans and specifications are reasonably accurate and complete for their purpose, design professionals, just like contractors, are human, and humans make mistakes!

The purpose of the study is to ensure coordination between the plans and specifications. It also has the benefit of identifying long lead items. The study is best carried out in the same sequence as the structure would be built. There is a natural "trigger" that prompts the project manager into asking: "What's the next step?"

Often in the heat of the bidding phase, questions are asked that require a written response that changes the bid documents. When that occurs, the architect or engineer will issue an *addendum*. Addenda (pl.) are sometimes hastily issued documents that can convey simple changes or have complex and far-reaching implications on the contract documents. Each addendum has to be reviewed in the context of the entire project and schedule. Lastly, it is helpful to ask the estimator his or her opinion on the completeness of the documents. The estimator may have already addressed many discrepancies during the bidding process.

During the bidding process the estimators learn a tremendous amount about the project. This is helpful information that should not go to waste now that the bid has been won. Estimators make assumptions during bidding. They ask questions and have conversations with subcontractor bidders. The estimators should share those assumptions, answers, and conversations with the project team. The best way is through a formal process in which the estimators share what they learned in the bidding phase. This gives the PM and team a head start in becoming familiar with their project.

Identify Discrepancies Early

The purpose of identifying discrepancies early is to avoid costly errors that can delay the progress or give rise to a claim or extra charge. Early detection allows the architect or engineers more time and a wider array of options to correct or clarify a discrepancy. Discrepancies that are discovered by the project manager before they become critical can be controlled and allow the project manager time to propose a solution before one is thrust upon him or her. The mere act of suggesting a solution, whether or not it is accepted, shows interest in the project. This has the added benefit of establishing the project manager as a partner or ally in the construction process instead of as an opponent.

Visit the Project Site

It's essential that the project manager take whatever time is necessary to visit and become familiar with the project site. Other team members such as assistant PMs and the superintendent should also visit the site. The site visit should *only* occur after the team has become familiar with the plans and specifications. The team must understand the physical aspects and uniqueness of the site *before* work commences. This is true about a new site, a renovation, or the addition to an existing building. Specific attributes such as utility pole location for power and telecommunications, water service connection points, and difficulty in accessing the site for large equipment or delivery vehicles are examples of the information that can be derived from a site visit. Natural features such as rock formations that may hinder excavation are another example. The existing condition of structures that are being added on to, or evidence of hazardous materials that are not identified in the documents, are all tangible benefits of site exploration. Again, the earlier the project manager takes the lead in learning about the actual site conditions and its limitations, the sooner that information can be disseminated to the rest of the project team.

Understand the Project Constraints

All projects have constraints. Probably the one that most construction professionals are familiar with is the time available for construction. Some constraints are contractual or imposed by the documents. Others can be a result of latent site conditions. Examples of contractual restraints include:

- Project phasing that prevents starting phase two until phase one has been completed and accepted.
- Occupied space that requires working around a business that is maintaining its operations.
- Utility changeovers that must occur during off-business hours.
- General context or process requirements,

Constraints that are a result of the site conditions include:

- Proximity to other structures or public roadways.
- Local ordinances that restrict construction activities during specific times or conditions.
- Lack of space for a staging or lay-down area.

Learning about some constraints is a result of the education that comes from the study of the contract documents, lessons-learned meeting with the estimating department, and possibly knowledge of local practice or ordinances.

Determine Potential Problem Areas

A natural extension of understanding project constraints is to look for areas that can pose a potential problem or delay. Every project has critical areas that can create a "bottleneck." These bottlenecks are identified by using the combined past experience of the team. This is where the old saying, "an ounce of prevention is worth a pound of cure" really applies. Potential problem areas such as the connection to existing water or other utility services and existing elevations at pipe inverts are but a few. The column line where the new addition is attached is also a good start. Existing conditions are rarely if ever as they appear on as-builts. A simple investigation or verification before the work has to be done can often save thousands of dollars and many days. As stated previously, early investigation allows the project manager to pose more cost-effective solutions or hire specialists that can supply solutions that prevent crisis or delay. At the very least, the PM is aware of where the high-risk tasks are and can act to mitigate the risk.

Understand Contract Procedures

Each project has contract procedures that are unique to the project and/or the owner: everything from payment schedules to administrative requirements. Others include insurance limits and indemnification clauses, inspection notification minimums, and updating intervals of the CPM schedule. Some less matter-of-fact procedures that the PM should understand include unfavorable contract language that assigns a disproportionate share of the risk to the contractor for specific events, no claim for delay clauses, time limits on claims notification, and dollar-driven CPM schedules that tie payment to performance. These are considerably less benign and should be in the forefront of the project manager's mind. While many of the contract procedures are identified in the Planning Process, they may not always filter down to the subcontractors and vendors. As these procedures apply to them as well, they should be aware of them and even built in to their agreements with the contractor.

Other more sophisticated procedures such as tax or energy credits/rebates by meeting certain predetermined objectives as having a project completed by a certain date can have far-reaching implications. All parties involved should be aware and held accountable in no uncertain terms.

The Schedule and Budget

As we will discover later in this text, two critical tools for project control are the schedule and the budget. Thus far, our focus has been on the schedule and its overall contribution to the project management process. It is the plan to accomplish the deliverables set against time. Equally as important is the budget. The budget is the dollars available to pay for the cost incurred as a result of performing the work to achieve the deliverables. The budget is a modified version of the estimate. It is the value of the work as seen through the eyes of the estimator, divided into elements that can be used to track cost in the performance of the work. Sometimes these costs are in the original categories in the estimate. Sometimes they are combined in *Work Breakdown Structures* or WBSs that represent the scope of a subcontractor's work. For each WBS there is a *Cost Breakdown Structure* or CBS that is the corresponding dollar amount available in the budget to pay for the work.

The schedule and the estimate are distinctly different but profoundly interrelated. It is the integration of schedule and budget that creates the baseline or performance metrics so critical to project control. These two topics are so important that entire chapters have been dedicated to each.

Summary and Key Points

This chapter reviewed the basic concept of project management with special focus on its application to the construction industry, also including a brief history of project management as a basis of project control and the contribution of two of its early practitioners. This chapter introduced the reader to several approaches to project management but highlighted the most common methodology and its five steps or processes: initiating, planning, executing, monitoring and controlling, and closing.

The benefits of proper planning are significant and provide an in-depth education on the project itself. As the lead on the project team, the project manager plays a role in the project management process that cannot be overstated. It is the project manager's duty to advance the goals of the project team.

Chapter 1 also explained the roles of the contract documents—plans, specifications, and addenda—in the project management process. It is one of the most important tools the project manager uses to manage.

The schedule and the budget are two critical components that create the baseline for project control.

Key points of this chapter are:

- Project management is the practice of planning, executing, monitoring and controlling, the work for a construction project.
- Project control is a subset of project management that focuses on cost and schedule as a means of determining performance.
- The traditional approach to project management often is the best methodology for construction projects due to its linear execution of tasks.
- The project manager is the team lead, often overseeing the development of both the schedule and the budget to be used as performance metrics.
- The project manager must understand his or her role and the contract documents. They must communicate this information in the development of the plan and the schedule.

Chapter 1 The Basics: Questions for Review

1. Name the two key principles described in Gantt's 1919 book, *Organizing for Work.*
2. Gantt was one of the first to recognize that in order for a team of workers to produce efficiently and to maintainable standards, what was required?
3. Fayol's book, *General and Industrial Management*, published in 1916, outlined a flexible theory of management that could be applied to most industries. What was the theme of the book?
4. Gantt is recognized for what visual tool which measures actual progress against planned progress?

5. Name the six management functions that Fayol is credited with that are still the basis of project management today.
6. Identify the five steps in the traditional approach to project management.
7. Name the five approaches to managing projects.
8. What is the most tangible benefit of the pre-construction Planning Process?
9. Define the components of the Monitoring and Controlling Process.
10. Explain the role of the contract documents in the control process.

CHAPTER 2

INTRODUCTION TO PROJECT CONTROL

Chapter 1 The Basics offered the reader a brief discussion of the concept of project management and how it is applied in the construction industry. Focus on the process was limited to what was needed for a frame of reference for later in this text. While project management is the parent discipline, its offspring, project control, is the focus of this text. Even a technically flawless project is soured by cost and schedule overruns and can damage relationships between parties. Therefore, project control is an essential aspect of the project management process. Whether the project is large or small, project controls offers the project team a snapshot of both cost and schedule performance that is essential to the decision-making process.

Overview of Project Control

Time and cost can easily get out of control on a construction project, even on a small project. While these two variables are independent, they are inextricably linked on all construction projects. Changes to schedule can affect the cost and vice versa. We have all heard horror stories of projects that ran 30 percent over budget and six months late in delivery. How can this possibly happen? In most cases, lack of or loss of control is the underlying factor. To avoid such a disaster, there needs to be a system for measuring the actual performance (in both time and cost parameters) against anticipated performance. The system for measuring progress has to be dynamic and ongoing during the life of the project. Waiting till the last minute to discover your project will deliver late is not only unacceptable, but offers little time (or hope) to recover. Equally important is the cost to deliver. Most construction contracts have a fixed price, or at the very least some form of accountability for the cost of the work as well as a time allotted for performance of the work. These two basic parameters create the foundation of project control.

Project Control: Integrating Cost and Schedule in Construction, Second Edition. Wayne J. Del Pico.

Enter the discipline of construction project control! Project control in its simplest terms is a four-step process of measuring progress toward a goal:

- Create a plan against which performance toward the goal can be measured.
- Formally and regularly measure progress toward the goal.
- Evaluate the causes of significant deviations from the plan.
- Take corrective action, based on the results of the evaluation, to bring the project back in line with the goal outlined in the plan.

All four of these steps are well within the duties of the project manager and team. The plan is the means and methods that the team adopts to execute the work, track costs and schedule weekly or bi-weekly, focus on those tasks that are not going according to the plan, and take action to fix the non-conforming tasks.

Project control is the function of integrating cost and schedule data to establish a baseline or guidance system for monitoring, measuring, and controlling performance. Project control can be performed by the project manager or for large projects can be an independent discipline within the project team performed by the *project analyst* or *project auditor*. Project control is the aspect of the project management process that provides the analytical tools for keeping the project on track, on time, and within budget.

Controls are instituted early in the project before the actual production tasks begin, and they are fully functional until the conclusion of the lessons-learned meeting and the archiving of the project files.

While some companies have universal control systems that are applied across every project, each project should be evaluated separately for the appropriate level of control required. Large, complex, one-of-a-kind projects with high risk generally require more control systems than small simple projects where the risk can be managed or loss to the business would be minimal. However, there are no hard and fast rules, each project should be evaluated on its own merit. There is one universal rule and that is that each and every project should have some level of project control. With too few controls, the project status is never really known. Too many controls can be cumbersome and costly to maintain, and also distract from the real project goals.

Control systems are typically established at a minimum for:

- Schedule
- Cost
- Contract modifications
- Risk
- Quality
- Resources

The *performance measurement baseline* is the time and cost parameters that the project team sets as the metrics to measure performance.

The performance measurement baseline (PMB) is the way the project would perform if it followed the planned schedule and planned budget exactly. The Monitoring Process begins only after the PMB has been established and agreed upon by the stakeholders.[1] The baseline values for both cost and time come from the budget and schedule, respectively. Since both are the contract values for the work, in theory, this is the maximum amount of money and time the contractor can expend to avoid a loss. These are the estimated or *planned* values for the task.

The gathering of information that will be used during the analysis portion of the control process is called *project performance measurement*. For starters, project performance measurement includes data on the cost of materials, labor, and equipment for work that has been performed or is being performed. It includes the individual productivities for the crew that performed the work and the duration of the task. These measurements are the *actual* values for cost and time to perform a task. This aspect of project control provides an integrated snapshot of both time and cost as of a specific date called the *reporting date*. This snapshot can then be used to discover trends and/or forecast future performance. It is used to analyze deviations from the planned performance called *variations*. If the variations are undesirable, the project manager can take corrective action to put the task or project back on track. If the performance variations are desirable, or, in other words, performance exceeds expectations, then the project manager can exploit whatever factor is creating the variation. Once the corrective action has been implemented, performance measurements can continue to verify that the action taken was appropriate and had the anticipated effect. Small adjustments to the corrective action can fine-tune the response so the team can get the exact desired response. This cycle plays over and over throughout the life of the project.

The project performance measurement is done on a micro scale. It requires that the project be decomposed into elemental components. This decomposition process is called the *Work Breakdown Structure* (WBS). The WBS elements are individual work packages, or groups of tasks, that can be monitored, tracked, and measured separately. They can be a task in the schedule or line items in the cost account.

Cost control provides the analytical procedures for monitoring, analyzing, forecasting, and controlling the costs on a construction project. Cost control requires a *Cost Breakdown Structure* (CBS) to match the WBS. It sets values for each element within a specific WBS. The cost control system is used to identify variations from the budget by comparing actual costs to the CBS. Cost control is most effective when integrated with project scheduling and progress measurement functions.

---

[1] In this case, the stakeholder can be any party involved in the tracking of cost or schedule depending on the contract delivery method.

Project Control Terminology

Throughout this book, we will frequently use project control terminology, so it might be beneficial to introduce the reader to some of the more common terms early. This list is not meant to be comprehensive; other terms and their definitions will be introduced as appropriate.

-A-

- **Accrued costs**, or **accruals**, are defined as the difference between the amount a project owes and what it has paid out. Accrued costs are recorded costs less amounts paid.
- **Activity** is a task that occurs in a schedule with a defined duration, start, and finish time. It is assigned to a resource to perform. It consumes time and money. It is used interchangeably with the term "task."
- **Actual Acceleration** is the request by the owner to accelerate the work to deliver the project ahead of the current Substantial Completion date. Actual acceleration typically adds cost to a project and is paid for by the owner.
- **Actual Cost of Work Completed (AC)** is the cumulative sum of the cost incurred in completing the work of a task, series of tasks, or the entire project.
- **Actual Cost of Work Performed (ACWP)** is the actual cost incurred in the performance of the work for a given time period. It is one of the principal measurement tools of project control.

-B-

- **Bar chart** is a graphic representation of activities depicted as a function of time.
- **Budget at Completion (BAC)** is the original budget plus any approved adjustments (change orders) at the completion of the project. It is the last Current Budget.
- **Budgeted Cost of Work Performed (BCWP)** is the value of the work completed measured in terms of the planned value of the work. It is the earned value for a task and a principal measurement in the Earned Value Management process.
- **Budgeted Cost of Work Scheduled (BCWS)** is the value of the work planned or scheduled to be accomplished within a specific time period, as illustrated in the schedule. It is one of the principal measurement tools of the project control process. It is often considered the baseline value.

-C-

- **Calendar day** is the measurement of the total schedule calculated in days against the calendar. It is the sum of work days and non-working days.
- **Cash flow** is the net flow of cash into or out of a project. It is a function of the project control process so that a company can plan for cash needs during the life of the project.
- **Change order** is a contract modification that changes scope, and possibly time and/or cost. Change orders can be added to or deducted from the contract sum. An approved additive change order represents an authorized addition to the current funding to pay for the change order. When a change order is approved, the current budget is updated by the amount of the change order.
- **Committed costs** are costs that the contractor is obligated to pay as soon as a purchase order is issued, or a contract signed. These costs become commitments to pay when the work has been performed. The difference between committed costs and accruals is timing. Accrued costs occur *after* the work has been performed, whereas committed costs are a pledge to pay as soon as the contract is signed.
- **Concurrent task** is one or more tasks that can occur simultaneously without impacting the other task(s).
- **Cost Breakdown Structure (CBS)** is the decomposition of the estimate into individual tasks or groups of tasks for use in the creation of the budget to measure performance.
- **Cost elements** are a subgroup of the detailed account. They are a further decomposition of the detail cost account into materials, labor, equipment, and labor-hours. Most job cost reports provide this level of detail.
- **Cost Performance Index (CPI)** is the budgeted cost of the work performed divided by the actual cost of the work performed. It is a ratio of earned value to the actual cost. Values greater than 1.0 are considered favorable; values less the 1.0 are unfavorable.
- **Cost Variance (CV)** is the difference (in dollars) between what was earned in performing the task(s) and what it actually costs to perform the task. Positive values are favorable and negative values are unfavorable.
- **Crash cost** is the cost associated with performing a task in crash time (or under acceleration).
- **Crash time** is the shortest time (duration) in which an activity can realistically be completed. It is measured in workdays.
- **Critical path** is the series of critical tasks that are linked in the schedule. A delay to the critical path will result in a delay to the Substantial

Completion date. It is the longest sequence of tasks that it takes to complete a project, but is the minimum project duration.

- **Critical Path Method (CPM)** is a scheduling method that calculates the longest path of planned activities to the finish of the project, and the earliest and latest that each activity can start and finish without making the project longer. It is based on the interdependency of tasks on the critical path.
- **Critical task** is a task on the critical path. Critical tasks have zero float and are dependent upon the performance of their predecessors to start on time.
- **Current Budget** is the Original Budget plus all approved change orders. The Current Budget reflects the current funding approved for the project. The Current Budget can and most likely will change over the life of the project.
- **Current Estimate at Completion (CEAC)** is an approximation or forecast of the total project cost at completion, as of the latest reporting period. This is based on updates to the current budget and committed costs, as of the latest reporting period.

-D-

- **Daily Output** is the amount of work calculated in units of measure (SF, LF, CY, etc.) specific to that task, which an individual or crew will produce in a workday (8-hours).
- **Deliverable** is a major goal, milestone, etc. that must be completed to finish a project, or series of tasks representing a phase.
- **Detail cost accounts** are the costs associated with a task or group of tasks that are being tracked as part of the cost control. It is the lowest level of the Work Breakdown Structure (WBS).
- **Direct costs** are those costs with a direct relationship to a task such as materials, labor, equipment, and subcontractors/vendors.
- **Duration** is the amount of time an activity takes to complete. In the construction industry, durations are measured in workdays of 8 hours.

-E-

- **Early Start (ES)** is the earliest date an activity can start. It is the latest early finish of its predecessor(s).
- **Early Finish (EF)** is the earliest date an activity can finish if all its preceding activities are finished by their early finish times.
- **Earned Value (EV)** is the value of work (in dollars or labor-hours) done to date either by line item on the Schedule of Values (SOV) or by the total SOV. It is calculated by measuring physical progress as a

percentage and multiplying it by the total value of the line item in the SOV. It is the BCWP.

- **Estimate** is a compilation of costs that represents the value of a scope of work as approximated by the estimator. It is the first step in project control and represents the original approved budget.
- **Estimate at Completion (EAC)** is an approximation or forecast of the total project cost when it is complete. This is similar to Current Estimate at Completion (CEAC).
- **Estimate at Completion Variance (EACV)** is the difference (in dollars) between the previous EAC forecast and the current EAC forecast. A positive value indicates the EAC has decreased. A negative value indicates the EAC has increased. It is used to indicate EAC trends of each reporting period.
- **Estimate to Complete (ETC)** is the approximate cost remaining to achieve completion. It is the EAC less total committed costs.
- **Event** is the point in time when an activity starts or finishes.

-F-

- **Finish-to-Finish relationship** is the relationship between two tasks on a schedule network in which their interdependence is based on their finish dates.
- **Finish-to-Start relationship** is the relationship between two tasks on a schedule network in which their interdependence is based on the finish of the predecessor and the start of the successor.
- **Float** is the amount of time in days between the time available to perform a task and the time required to perform the task. Float is measured in calendar days. Tasks with a float of zero days are defined as critical tasks and are on the critical path.
- **Fragnet** is a small portion or fragment of a schedule network.
- **Free Float (FF)** is the maximum amount of time a task can be delayed from its Early Start (ES) without delaying the ES of any activity immediately following it.
- **Forced Acceleration** is a directive by the owner to the prime contractor or from the prime to a subcontractor to accelerate the work when the work is behind through no fault of the owner. Forced acceleration is typically not compensable.

-G-

- **Gantt chart** is a type of bar chart or schedule that illustrates tasks as a function of time. Gantt charts do not always show interdependency between tasks. Gantt charts are named after Henry Gantt.

-H-

- **Hard logic** is the relationship between two tasks on a schedule that is based on a physical relationship as opposed to a constraint (e.g., the formwork has to be erected before you can place the concrete in the form).

-I-

- **Independent float** is the amount of scheduling flexibility available on a task without affecting any other task.
- **Interdependence** is the dependency relationship between tasks on any path.
- **Indirect costs** are those costs that are nonproduction costs or overhead costs. There is no direct relationship to a specific task but to the project as a whole. Examples include home office overhead costs that are absorbed by all projects.

-J-

- **Job cost report** is a printed cost report generated at regular intervals showing costs incurred by detail cost account and cost element. It is compared with the current budget for the same cost account. It is computer-generated as an accounting/bookkeeping function.

-K-

- **Kickoff meeting** is typically the first meeting with all team members for a project. It often sets the tone and expectations of the various parties for the project.

-L-

- **Lag** is the amount of time in days a dependent task must be delayed before it can begin or end.
- **Late Finish (LF)** is the latest time an activity can finish without delaying a succeeding activity.
- **Late Start (LS)** is the latest time an activity can start without causing delay to a succeeding activity. It is the latest finish of all activities preceding it.
- **Learning curve** is a mathematical curve used to predict the reduction in time resulting from a task being performed repeatedly. Once the learning curve has been expended, the task should be performed at maximum productivity.

-M-

- **Management Reserve** is a designated amount of money set aside in the owner's budget to account for costs of the project that cannot be predicted in the planning stage. It is not accessible by the contractor and is sometimes called the owner's contingency.

- **Merge Activity** is an activity with more than one predecessor immediately preceding it. Merge activities impact the start of the succeeding activity.
- **Milestone** is a special event that marks the completion of a task, phase, or portion of work. Milestones mark an accomplishment on the schedule. They act as way markers to measure the progress of the project as incremental portions of the work get accomplished. Milestones are on the critical path of the schedule.
- **Mitigating risk** is an action taken to reduce the likelihood of a risk occurring or an action taken to reduce the impact that a risk will have if it does occur.
- **Monte Carlo Simulation** is a method of simulating task durations using probabilities evaluated by a computer program. The method identifies the percentage of times, tasks, and paths that are critical over numerous simulations.

-N-

- **Network** is a logic diagram arranged in a predetermined format to illustrate the tasks, sequencing, dependencies, and relationships between specific tasks on a schedule.

-O-

- **Original Budget** is established once the contract has been awarded. It is the contract sum less the included profit. It is the sum of all anticipated costs applicable to the project. The Original Budget for a project does not change.
- **Over-allocated resource** is a resource—labor or equipment—that has been overscheduled based on its availability.
- **Overhead costs** are the costs associated with running a business or a project. It includes indirect overhead costs such as management salaries, home office expenses, and other overhead costs directly related to a specific project.

-P-

- **Path** is a series of interconnected or interdependent tasks that are critical or concurrent.
- **Performance Measurement Baseline (PMB)** is the schedule and budget values that the project team agrees upon as the standard for comparison with actual progress and cost.
- **Physical progress** is the quantitative measurement of the amount of work that has been completed. It is typically represented as a

percentage of the total. That percentage is multiplied by the line item value in the SOV to arrive at the earned value. It is an important measure of how a task or project is progressing.

- **Planned Value (PV)** is the estimated value of work to be performed within a given duration or period. It is used to create the performance measurement baseline. It's also known as the BCWS.
- **Productivity** is a measure of the efficiency of production. It is the ratio of what is produced to what is planned to produce. It is typically measured in labor-hours (or dollars). It is defined as the earned value (in labor-hours) divided by the actual labor-hours expended. Values greater than 1.0 indicate that work is progressing faster than planned. Values less than 1.0 indicate that more resources have been expended than planned.
- **Progress payment** is the amount invoiced by the contractor for work completed in that period (typically monthly). It is the summation of the earned value of each line item in the Schedule of Values less any retainage.
- **Project closeout** is the final phase of the project, beginning at Substantial Completion, in which the closeout documents—manuals, technical information, waivers, as-builts, etc.—are turned over to the owner in a formal process. Project closeout typically ends with the lessons-learned meeting and the archiving of the project files.
- **Project manager** is the lead on the project team responsible for the management of the project from inception to closeout.
- **Project Performance Measurement (PPM)** is the gathering of cost and schedule information to be used as a comparison to the Performance Measurement Baseline.
- **Project team** is the group of individuals responsible for the planning, execution, monitoring, and control of a project.

-Q-

- **Quality** is the particular characteristics, attributes, or nature of a product, execution, or installation of a task on a construction project as defined by the contract documents.
- **Quantity survey** is referred to as the takeoff. It is the decomposition of the project into tasks with quantities at the beginning of the estimating process. The quantities are multiplied by the unit price costs of the task and extended to the total.

-R-

- **Recorded costs** are costs that have been incurred during the construction process for elements such as materials, labor, equipment, subs, and overhead. Recorded costs are represented by an invoice and are tracked by the accounting department in most construction companies.

- **Recovery schedule** is a schedule illustrating the logic and sequencing of tasks to put the project back on track after it has slipped from its original schedule.
- **Resource leveling** is a technique used to reduce the demand for resources by scheduling them more efficiently and using float to prevent over-allocation of a resource.
- **Retainage** is the cumulative amount withheld from each progress payment until the completion of the project. It is to ensure completion of the punchlist and warranty items. Retainage is traditionally the final payment on the contract.
- **Risk allocation** is the sharing of risk with other parties. Traditional methods include subcontracting of scopes of work that the contractor lacks proficiency in or labor resources to complete.
- **Risk avoidance** is the strategy by which a business decision is made not to bid a project due to a variety of risk factors such as ambiguous or incomplete contract documents, a difficult owner, an unrealistic schedule, or some other high-risk factor.
- **Risk profile** is a series of questions designed to reveal specific areas of uncertainty in a project.
- **Risk retention** is the result of an economic evaluation of the loss exposure and the determination that the loss can be absorbed safely. It is also the policy of retaining the risk to maintain absolute control of the task that exposes the contractor to the risk.
- **Risk transfer** is the shifting of risk from one party to another by the use of insurance, bonds, or indemnification statements. Subcontracting can also be considered risk transference.

-S-

- **Scenario analysis** is a process in which potential risk events are identified and analyzed to determine likelihood and impact.
- **Schedule of Values (SOV)** is a detailed allocation of the contract sum into individual components or categories of work. The Schedule of Values is the basis for determining earned value for progress payments during the life of the project. The scheduled value of a line item is the total value of that line item at 100 percent complete.
- **Schedule Performance Index (SPI)** is the budgeted cost of work performed (BCWP) divided by the budgeted cost of work scheduled (BCWS). It is a ratio of the earned value to the planned value. Values greater than 1.0 are considered favorable; values less than 1.0 are unfavorable.

- **Schedule Variance (SV)** is the difference (in dollars) between what was planned on being earned and what was actually earned for a task. Positive values are favorable and negative values are unfavorable.
- **Slope** is the ratio of cost per day of duration to shorten or crash a task.
- **Soft logic** is the relationship between two tasks on a schedule that is based on a constraint as opposed to a physical relationship (e.g., the task has to be rescheduled due to the unavailability of a crew or piece of equipment to perform the task).
- **Splitting** is a scheduling technique in which the work on a task is temporarily interrupted so that the resource can be assigned to another task, then reassigned to the original task to complete.
- **Substantial Completion** is the date on which the owner/end user can take beneficial occupancy of the project for its intended purpose. It is traditionally the date on which the warranty period begins.
- **Summary cost accounts** is a summation by group of detail cost accounts that comprise a high-level Work Breakdown Structure (WBS). It provides a "big picture" view of a particular WBS.

-T-

- **Task** is an item of work that needs to be done. It consumes resources and has duration. It is also called an activity.
- **Total Allocated Budget (TAB)** is the owner's Current Budget plus the Management Reserve.
- **Total committed costs** are all recorded costs, including accruals, plus any committed costs yet to be recorded.
- **Total float** is the amount of time a task can be delayed without delaying Substantial Completion, a constraint, or the end of the schedule.
- **Tracking Gantt** is a Gantt chart that compares actual progress to planned progress.

-U-

- **Unilateral** is a change or action taken by or involving one side or party only. It most often refers to an owner's contractual right to request and issue change orders during the life of the project.

-V-

- **Variance** is a difference from what was planned or anticipated. Most often used in addressing the difference in budget or schedule from what was planned.
- **Variance at Completion (VAC)** is the Budget at Completion less the Estimate at Completion (EAC) or the total amount budgeted less the total cost of the project. Positive values are favorable and negative values are unfavorable.

-W-

- **Work Breakdown Structure (WBS)** is a decomposition of the work into smaller or more detailed components. The WBS has multiple levels. The lowest level of the WBS is the detail cost account. It is used to allocate costs from the SOV.
- **Work package** is a task at the lowest level of the WBS.

Project Control Formulas

In addition to terminology, the following are the formulas most commonly used in project control:

- Cost Variance: CV = BCWP – ACWP
- Current Budget: Original Budget + Approved Change Orders
- Schedule Variance: SV = BCWP – BCWS
- Variance at Completion: BAC – EAC
- Cost Performance Index: CPI = BCWP ÷ ACWP
- Schedule Performance Index: SPI = BCWP ÷ BCWS
- $PERT_{Mean} = [O + 4(ML) + P] \div 6$
- $PERT_{Standard\ Deviation} = (P - O) \div 6$

The Project Control Cycle

For the purpose of this text, the term *life cycle* can be loosely defined as the series of stages that a construction project goes through from beginning to end that characterizes that project. There are two life cycles that work in unison with one another in every construction project. The *project life cycle* is the series of tasks required to produce the deliverable: a building, bridge, road, etc. Project life cycles are very different for different types of structures.

The *project management life cycle* defines the steps needed to manage a project through the project life cycle (the five steps of the Traditional Approach introduced in Chapter 1). It is reasonably the same for every construction project.

Within the project management life cycle is the *project control cycle.* The project control cycle is defined as the series of steps required to monitor and control the costs and schedule of the project.

The project control cycle is as follows:

- Plan how the project will achieve its goals.
- Execute the work according to the plan and schedule.
- Monitor the progress of the plan.
- Measure the progress of the plan.
- Identify variations from the plan.

- Analyze cause of the variations.
- Implement corrective action(s) to realign the progress (or cost) with the plan.
- Execute the work after the corrective action has been implemented.
- Measure changes to the progress as a result of corrective action.

Figure 2.1 illustrates the cycle of project control for a construction project. Each of these features of the project control cycle will now be explained in the order that they proceed.

Plan to Achieve Goals

Prior to the commencement of any production work on-site, the contractor's project team decides how the project will be built: the "means and methods" portion of the plan. The expertise of the team is used to create the most efficient use of time, money, and resources to get the project to Substantial Completion. The team evaluates those tasks or stages in the project that present the most risk to achieving the project goals. The team considers options to mitigate the risk or transfer the risk to other participants. Options may include performance and payment bonds or the decision to subcontract a task versus self-performing the work. The team studies the contract documents and analyzes the estimate. They proceed to decompose the project

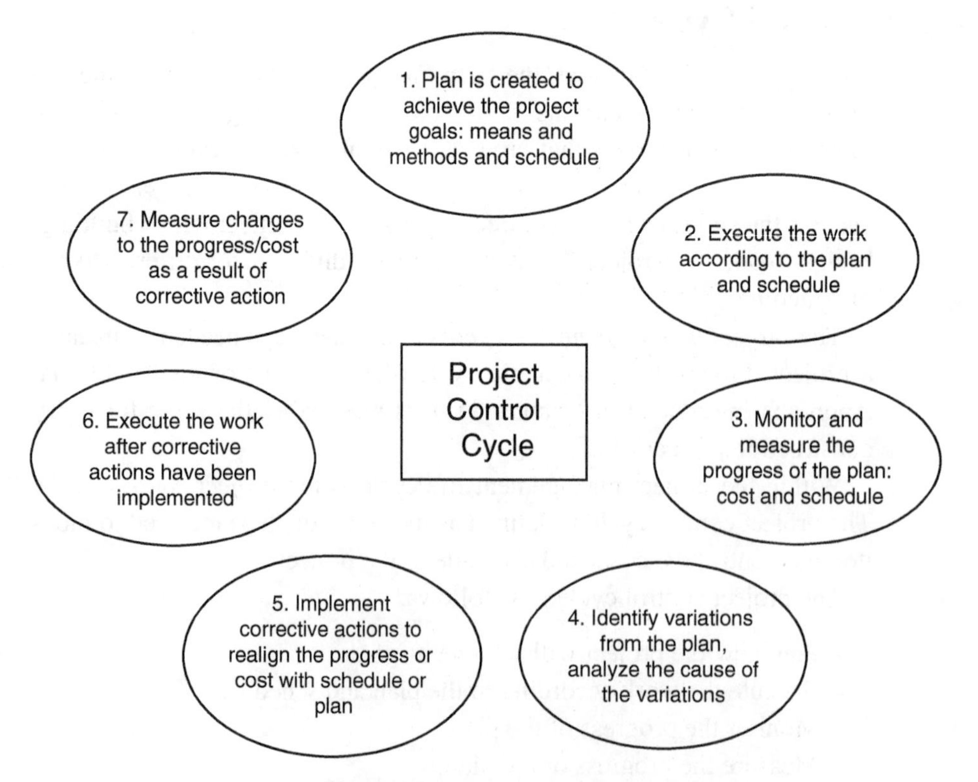

Figure 2.1 Project Control Cycle Diagram

into its component tasks and establish the work breakdown structures that will be used in both the schedule development and the detailed cost accounts. These will be the basis of the performance measurement baseline (PMB), or the metrics by which the team will measure performance. The team meets with participants (subcontractors, vendors, suppliers, etc.) and solicits opinions to gain insight and contributions to the plan and schedule, as well as the essential "buy-in." They evaluate the appropriate resource for the task and decide on required labor levels. The schedule is developed and revised, adjusted and revised again, and after several generations of schedule, it is ready to become the chief project management tool. Costs are loaded against tasks in the final schedule and the cost accounts in the job costs system receive their "estimated values" based on the decomposition of the estimate. The project team makes final decisions on internal procedures and who will represent the project in the field. Contracts are signed, purchase orders are issued, and commitments are made, all in the furtherance of the project goals. The planning phase is complete.

Execute Work According to Plan

The kickoff meeting is held, the communication chain is established, and all participants understand the project expectations. The project is mobilized. The schedule is reviewed to ensure each participant is aware of the performance expectations in time and magnitude. The production process starts and both employees and subcontractors get acclimated to the project and the schedule. Learning curves are mastered.

Regular meetings (most often weekly) are held to discuss progress in comparison to the schedule. As the project "ramps up," portions of the plan and schedule may have to be adjusted for on-site realities. This is not uncommon. Remember the schedule is dynamic; it is meant to change!

Those performing in accordance with the plan and schedule are routinely passed over as "doing their job," while the real focus is on those tasks or participants that are behind on the critical path. Concerns are documented by the on-site managers along with any apparent reason for the lack of production. As part of the feedback process, this information is passed up to the project manager. If necessary, additional team members and even senior management weigh in, based upon their area of expertise.

Measure the Progress

The measurement phase evaluates actual performance and compares it with planned or anticipated performance. For those tasks determined to be behind, the project manager and team members monitor the performance of the task for anything that may be immediately evident as the cause of the reduced productivity. Quite often, reduced production can be attributable to learning curves. Causes are numerous and varied: lack of materials, insufficient instruction, incorrect tools, sequencing, crew size, and weather. For subcontractors it can also be that they have other commitments. Details of the crew composition, work habits, sequence of operations, and daily outputs are recorded and analyzed.

Identify Variations

With the data in hand, deviations from the planned outputs are identified. Is the reduction in productivity within expected tolerances? Is the reduction detrimental to overall project schedule? Can work be resequenced to reverse the loss of time?

Analyze Cause of Variations

Each variation from the planned performance has a cause or even multiple causes. Determining the factor that is the culprit may not always be a simple task. The process of analyzing the variation should start with the simplest possible cause working up to the more complex possibilities. Simple causes often require simple fixes and simple fixes cost less. Case in point: Asking a simple question of the foreman—"What would make this process easier and faster?"—might reveal that something as simple as moving the staging area closer to the work would save cycle time from the lumber stockpile to the work area. While this may be an oversimplification, some solutions are very simple and may be handled at the field level.

Implement Corrective Actions

Once the cause becomes apparent, the project manager must take action. While hoping things will improve and praying for divine intervention are both positive, neither one is a professional course of action. Take the most cost-effective course of action to correct the problem as soon as practically possible, but only after all of the facts are carefully reviewed. While hasty, emotional, or uninformed decisions are not the course of action of the professional, waiting only exacerbates the problem.

If the corrective action requires consensus of the project team, call for a meeting or send out an email. If the fix requires the approval of the upper management or the client, get the approval as soon as possible.

A word of caution: Seeking approval outside the team, while not uncommon, should not be standard procedure for all corrective actions. It should be only for the more complex, costly fixes. Continual delays while waiting for a decision from the "board" hampers a PM's ability to be proactive. On a similar note, solutions that require approvals from outside the team should be less experimental and have a higher expectation of success. Too many failed attempts at correcting the same problem can result in the PM's decision-making ability being called into question.

Execute Work and Measure Changes

The task should continue after the corrective action has been implemented. Once again, tasks are monitored, performance is measured for improved productivity, and the success or failure of the corrective action is documented. The documentation is not only as a means of providing recorded closure to the problem, if successful, but as a possible solution to a future problem of a similar nature. These are fundamental exercises that can add benefit to a lessons-learned meeting at the end of the project.

Not all initial corrective actions are the right remedy to the problem. Occasionally, multiple actions have to be tried before the correct solution is found. This should not reflect poorly on the PM or the team. After realizing the solution is not correct, reanalyzing the data or reviewing the new data and trying another corrective measure is part of the process! Sometimes the initial corrective actions need to be adjusted based on new information or in light of how it actually works.

Summary and Key Points

In the overview section of this chapter, the reader was introduced to the project control process by identifying where project control fits in the overall project management profession and who performs project control.

The two main components for project control are time and cost, which are represented in real project terms by the budget and the schedule. It is from the budget and schedule that the performance measurement baseline is created and set in place as the guidance system during the project life cycle. It is the integration of the budget with the progress of the schedule that allows the project manager to know exactly where the progress of the project lies.

The reader was introduced to some of the more common terms used in project control and a step-by-step elaboration of how the project control cycle works during an average construction project.

Key points of this chapter are:

- Cost and schedule are used together to determine project status.
- Project control starts with the creation of a plan against which one can measure performance.
- Formally and regularly measure progress toward the completion of the plan.
- Evaluate performance as measured against the plan.
- Take corrective action as required to realign with the plan.

Chapter 2 Introduction to Project Control: Questions for Review

1. Project control is the function of integrating cost and schedule data to establish a baseline or guidance system for monitoring, measuring, and controlling performance. True or False?
2. Control systems are established for which six factors in the management process?
3. The gathering of information that will be used during the analysis portion of the control process is called?
4. What is the name of the time and cost parameters that the project team sets as the metrics to measure performance?
5. Deviations from planned performance are called?
6. What is the name of the decomposition process that breaks the project down into its element components of work?
7. Identify the acronym ACWP and provide a definition.
8. Identify the acronym BCWS and provide a definition.
9. Identify the acronym BCWP and provide a definition.
10. Define the formula for the Schedule Performance Index (SPI) and explain what it is a measure of.

CHAPTER 3

PRE-CONSTRUCTION PLANNING

Chapter 2 Introduction to Project Control outlined for the reader how the project control process works, where it fits in the project management process, and some of the more common project control terms and formulae. Now that the reader has a general understanding of the process and why it exists, the next step would be to discuss how it is initiated. This chapter will focus on the planning stages and setup for project control.

Initiating the Control Process

As with any business, managers have a constant need to know how the business is performing, not only from a time standpoint, but also from a cost perspective. Most for-profit companies, whether publicly traded or privately held, have investors. Investors can be outsiders to the company that provide capital through the sale of shares or stock, or the investors can be the partners that own and operate the business that have to reach into their personal assets to build the business. In either scenario, investors contribute money in the expectation of a return for their investment. That return comes from the profit earned. In order for investors to do their part, they expect that managers and the individuals who operate the business will take concrete steps to reduce the risk assumed by the investors.

One of the best ways to reduce risk is to understand where it occurs in each project and to define it in measurable terms so that the responsible managers of the project can create safeguards to minimize the risk. While there are numerous and differing risks with each construction undertaking, the two most common and significant are the risk of failing to complete on time and a negative return for the money invested.

Creating safeguards to minimize risk and metrics for measuring how successful we are at performing the work is the essence of planning. Planning is done in advance of the actual performance of the work. In our industry it is called *Pre-Construction Planning*, and, at the risk of using an overworked cliché, it is planning for success!

Project Control: Integrating Cost and Schedule in Construction, Second Edition. Wayne J. Del Pico.

Pre-Construction Planning

Planning in general can best be described as the function of identifying the goals of the project and then establishing the policies, procedures, and practices for achieving those goals. Planning in the construction industry can be described as selecting a course of action with a predetermined outcome. Planning is also decision making since it involves deciding between alternatives and options.

As stated earlier, established successful construction companies have policies and procedures for managing a project. Policies and procedures are designed to control an action in order to provide a predictable outcome. While managers all have different managing styles, they are all required to follow the same policies and procedures. Imagine the confusion if project managers were allowed to record and track costs as they chose instead of following company protocol. Efficiency and consistency would be cast aside. Each team member would be required to learn the idiosyncrasies and mechanical processes of the individual project manager. Clearly, this is not the best avenue for overall success.

Some policies and procedures are shared throughout the industry and some will vary from company to company. The intent is to streamline the process of managing the routine portions of a project and to make it as efficient and effective as possible. If we can reduce the time spent on routine (but necessary) tasks, it leaves more time for the analyzing and problem-solving aspects of the process. The key to efficiency and effectiveness is to ensure that all team members are aware of the proper procedure for specific processes within the overall management of the project. Communicating the established procedures and company policy for managing the project is paramount. Policies as simple as setting and maintaining an updated schedule to inform users that the data is the latest available, payroll hits the job cost report at noon on Wednesdays, or replacing a folder in the central paper file with a card indicating who has it, can save time and possible revisions to completed work. While simple, they can be extremely effective in operating professionally. Policies and procedures set the functional guidelines of the team members for performing their duties. They become used to the regiment and availability of information. Even for those team members who are not in-house, establishing chains of communication can save hours of wasted time and delayed or missed opportunities.

The pre-construction planning phase is intended to decompose the project from a different perspective than the estimator had. It is intended to evaluate the constructability of the project and the documents that will be used as the guidance system for that process. The greatest asset of the planning phase is that it "forces" the participants to think in detail and consider a step-by-step analysis of how the team will build this project.

The planning phase is often quite revealing about anticipated or previously held means or methods for executing the work. The greatest benefit of this phase is that process by which the team determines what will not work, what will work, and what will work best. The construction industry has it relatively easy in comparison to some other industries in that by the time the pre-construction planning phase has started, most, if not all, of the two largest components—estimating and scheduling—have been established. The project has been bid, and as part of that bid, time for construction has been established or the bidder has accepted the owner's timetable. Another major aspect of the planning process that construction as an industry avoids is the defining of the project. Our projects are defined by the plans, specifications, and the contract. These govern what we will build and how we will build it. . . to a point.

While the plan agreed upon is never carved in stone, it is meant to evaluate the best way to accomplish the goals, and in the event the "best" way turns out to have been less than the best, it allows the team to consider alternatives or a Plan B in advance of needing it.

The Planning Process also identifies and communicates project priorities. These priorities can be of a contractual nature or can be imposed by the team. By identifying the priorities, the team can place a value on each priority which allows the project managers to make appropriate trade-off decisions if needed later in the process.

The Planning Process also establishes the *communication plan*. Every project manager has a recollection of a project that suffered from failure to communicate. The communication plan is a key component in coordinating and tracking schedules, issues, and action items. It is most often created early in the planning phase by the project manager with the help of the team. A well-defined communication plan can alleviate problems and ensure that all team members and stakeholders have the information needed to do their jobs. Good communication skills are learned, they are not a genetic trait. Establishing parameters and regular procedures for communication is a good start to learning the skills needed.

Pre-construction planning, similar to estimating and scheduling, is done by professionals experienced in the type of project and the means and methods that will be employed to accomplish the work. Team members should be selected for their skill sets and demonstrated areas of expertise, not because they were the next available project manager or superintendent.

Keeping the planning process on target can be difficult. Some suggestions for moving forward include:

- Identify the goals at the very beginning and ensure all participants understand the goals.
- Initially, let team members do their own independent planning.

- Encourage "big picture" planning to start, with details to be refined later.
- Remain flexible and open to (almost) all ideas.
- Encourage participants to look for the flaws—constructively!
- Try to test the assumptions behind the ideas.
- Reach outside the team for expertise that is not available in-house.
- Involve upper management whenever practical. Remember that they may have to bless the plan and their contributing goes a long way toward approval.
- Keep in mind that the more complex the idea, the greater risk involved.

Insight that has been gained in lessons-learned meetings from similar projects can save time and money, and avoid repeating costly errors. These same meetings reveal which team members and their skill sets were successful on the last project that was similar to the current one. The individuals on the project team acquire significant learning during the pre-construction planning, and this education enhances their ability to manage the project.

Not all team members can be handpicked; certainly, those within the owner's or design professional's group are outside the project manager's purview to select. In this situation each team member should identify their responsibilities so there are no misconceptions or expectations that are not met. This is often done at the project kick-off meeting. This can also be a good place to establish the communication channels that the project will follow.

Understanding the responsibilities and contributions of each team member may be a helpful start in knowing what to expect and how to assemble the appropriate team within the project manager's control. Again, the project manager has little control over those team members in the employ of others.

Key Personnel and Responsibilities

In Chapters 1 and 2, The Basics and Introduction to Project Control, much has been mentioned about the project team and their collective responsibilities. Let's analyze the project team and discuss responsibilities and contributions of the individual team members. The following is meant to illustrate possible team members, but is in no way a comprehensive list. The team members have been organized in the groups in which they would normally occur.

The Contractor's Group

The contractor's group is charged with the execution of the work in accordance with the terms of the contract. They are the constructor and should be the driving force behind the pre-construction planning phase. The management portion of the contractor's group consists predominantly of in-house employees and, to a lesser degree, subcontractors and vendors that will have a continual interface with the project. The following, in no specific order, are the contributions to the team and the project control process of the project manager, estimator, scheduler, superintendent, accounting staff, administrative staff, and subcontractors/vendors/suppliers.

Project Manager

The project manager (PM) is the lead on the project team. He or she is the decision maker in the feedback cycle who is responsible for overall project performance: success or failure. PMs report to senior management and are responsible for all team members receiving regular updates on project status. In many companies, they are also the project scheduler and the project control's analyst. The PM works with the estimator and other team members to arrive at the detailed cost accounts that will be the basis of the project cost control. The PM assembles the periodic billing, reviews it with the architect or owner's rep, and sees that it is approved and paid. The PM also approves or rejects payments to subs, vendors, and suppliers. Any value engineering initiatives, while not always initiated by the PM, are directed by the PM. The project manager chairs the meetings in the office and on-site. With the contributions of team members, the PM sets the meeting agenda to ensure that pertinent issues are discussed and resolved in a timely manner.

PMs are highly organized and capable people managers. They are experienced and possess a sound technical background and solid communications skills: both written and verbal. They solicit information from the appropriate team member experienced in the issue at hand in order to make a decision that furthers the project goals. As seasoned professionals, they recognize that no single individual can do it all, and so they recognize other team members' contributions to the planning and management.

Estimator

The estimator is key in educating the project team as to how the estimate was prepared and ultimately how the project was bid. This is not limited to costs, but also the means and methods, which are the details behind how the project is to be built and the basis of how the project was estimated. The estimator is charged with ensuring that the project team knows how the project was estimated. Some examples of information conveyed by the estimator include: Were there negotiations that led to the securing of the contract? Was a portion of the work estimated to be fabricated off-site to reduce costs and improve quality? Was the contract awarded based on value engineering proposals? Did the bid contain approved material substitutions? Why was that subcontractor carried in the bid? All of these fine points that were the basis of the estimate have a tremendous impact on the process and therefore the budget that will be used for the baseline of the control.

It should come as no surprise that schedule can have a tremendous impact on a project award as well. In fact, on some projects, it surpasses price as the main criterion for award. The estimate is comprised of both fixed costs and time-sensitive costs. Many of the costs associated with project overhead are time-sensitive, which is to say that they are directly affected by the schedule. Consider the project superintendent: His or her salary is based on the amount of time spent on-site. Therefore, a project that has a 52-week schedule will have 52 weeks of superintendent's salary associated with it in the estimate. A 26-week project will have less time-sensitive costs in the project overhead. Some other examples of time-sensitive costs include: office and storage trailer rental; temporary facilities such as toilets, lighting, and power for the site or trailer; and site administrative staff. Examples of a fixed cost would be a building permit. Pay for it once and that's it, regardless of the schedule of the project.

Does that mean that a schedule is developed at bid time? You bet! While most projects provide the bidder with a definitive time available for performance of the work, this is not the same as a schedule. Consider a project with a 270-day performance time, which is to say the specifications call for the project to be Substantially Complete 270 days after the Notice to Proceed is issued. This is a contractual requirement imposed by the owner through the documents. This is the maximum duration allowed under the contract, not necessarily the schedule. As determined by the estimator, and despite multiple revisions and scenarios, the schedule-at-bid does not exceed 240 days, would it not be more prudent to carry 240 days in time-sensitive costs to gain a competitive advantage? The answer is a resounding yes! The inverse should also be considered. If the schedule-at-bid cannot be brought in line with the 270-day window and despite what means and methods are employed, the schedule is 300 days, the team has to consider what risks are posed by this dilemma. Should the estimator add overtime hours and costs to accelerate the work?

Consider colder, northern climates where the seasons have a direct impact on performance, as well as means and methods: Without a schedule, how does the estimator calculate how long winter protection will be required? How would the estimator know whether the exterior masonry starts in October or January? These are critical pieces of the puzzle and need-to-know information for the management team.

All of this information must be conveyed to the project team. While the estimator may have a somewhat limited role after the project is awarded, their handoff of the project to the team that will fulfill the contract is extremely important to the development of the plan and schedule. This conveyance of information is done at a meeting (or meetings) where the estimator walks through the estimate with the team. Conveying the intricacies of the estimate is essential for creating a solid foundation for the planning and scheduling phase.

The estimator may also work with the project manager to create the Work Breakdown Structure (WBS) and assign initial planned values to each cost element in the detailed cost account. These are key to comparing actual to estimated costs that are part of the baseline and become the cost breakdown structure.

Scheduler

If the company has a scheduler, that individual may have been involved in the development of the preliminary schedule used when the project was bid. This can often be an advantage to the planning team. It is a place to start or at least to understand the logic behind the schedule used when the project was bid. This is not to say that this is the final or only version of the schedule the team will use. It is, however, the schedule or, more importantly, the project duration that constitutes the basis of the time-sensitive project overhead costs.

The scheduler contributes to constructing the WBS that will make up the project schedule. From the WBS the scheduler helps the team decompose the project into activities (tasks) and identify major milestones that are key to the performance measurement baseline and the contract. The scheduler helps in answering one of the most difficult questions: "What is the right level of detail for the project schedule?" Too much detail and the schedule is intimidating; too little detail and it fails as a management tool.

The scheduler determines the duration of the tasks to be used in the project schedule. This is done through a series of calculations based on quantities, crew sizes, and daily outputs as well as contributions from major subcontractors and vendors. The project team, with the help of the scheduler, establishes the appropriate milestones for the schedule.

Superintendent

As the on-site representative for the contractor, the superintendent, or *super*, is responsible for the day-to-day operations. He or she ensures that the work is executed in accordance with the agreed-upon plan and the schedule. The super is the eyes and ears on the ground, and is the author of the Daily Report. The Daily Report is the main field-to-office written communication tool. It is a key component in the feedback cycle that is so important to the control process. Its value cannot be overstated. It allows the PM to measure progress and decide if on-site performance is meeting planned values. The Daily Report is an essential source of information that is used in decision making. The super analyzes daily outputs for a crew and keeps running totals on production. Many superintendents have a "feel" for progress based on experience. He or she compares actual outputs to anticipated outputs of the various crews to determine if the task will complete early, on time, or late. The super can then take the appropriate action or inform the project manager. He or she typically has a technical background and considerable experience in the type of construction under contract. The super should be organized and have one foot in today and the other foot in tomorrow. This one-step-ahead paradigm allows supers to plan for tomorrow, next week, or three weeks down the road.

Accounting Staff

The accounting staff assists the project team in assigning the appropriate cost codes to build the detail cost account. This is a pro-active position on the team. The cost elements complement the WBS structure that make up the tasks of the schedule. They input the initial planned values for each cost element that is provided by the project manager. These planned values serve as the baseline for cost comparisons that are so crucial to the financial control aspect of project control. Once the detail cost accounts have been set up, reviewed, and approved by the PM, the *tracking* or recording of funds expended can begin.

The accounting staff plays the important role of tracking costs approved by the PM. Invoices from vendors, suppliers, and subcontractors (recorded costs) for materials supplied, equipment used, or work performed as well as routine invoices for project overhead costs are reviewed, approved, and coded. All costs are coded or assigned by the PM to specific detail cost accounts and then the cost elements within. These invoices are *posted* or entered in the computer along with weekly payroll, all of which results in a computerized report that identifies exactly where the project is financially as of a specific date.

The accounting staff is often the first to see the job cost report. This allows the financial officer to get a "big picture" view of the financial status of the cost side of the project. Although the financial officer and the staff may not be fully privy to the reasons why costs are over or under planned values, they can alert the project manager to variances and even trends that are cause for concern. The accounting staff can also compare receivables to payables to ensure that the financial model for the project is being achieved and the project is not "upside down."

A cooperative financial officer and accounting staff can be an enormous boon to the project team. Conversely, an accounting staff that does not track costs in a timely manner prevents the project team or senior management from ever really getting an accurate picture of the financial status of a project in real time. Job cost reports are a routine communication tool. They should be available to the team at specific intervals determined by the team in advance.

Administrative Staff

The administrative staff, or *admin*, is often the workhorse of the team. They are responsible for the updating of submittal, RFI (Request for Information), and change order logs, among others. They follow up with team members to ensure that they have what they need for meetings. Administrative assistants are keepers of the project record: all of the documents that create the history of the project. Without regular and comprehensive updating, the team is without crucial information needed for the decision-making process. The admin may be the first to know if a change order has been approved or an RFI has been answered. They are charged with routine communications such as meeting notices, reminders, and the distribution of meeting minutes, to name but a few. For large projects whose budget can support the cost, an assistant PM will assume some of these functions. While the admins traditionally play a diminished role in the decision making, their contribution is essential to the project team. A good admin is an incredible asset to the team.

Senior Management

The role of senior management for the contractor is often understated. They are intended to serve in most cases in a less active, more passive advisory role. They are there for project managers to turn to for advice and consent, or to simply bounce an idea off. Securing senior managements backing for a change to the plan, especially one that costs money, can often define the line between success and failure. Most PMs learn very early in their career that overtime is not authorized without the approval of senior management of the company.

Senior management of the company also has the big picture view of the project, they may have set the ancillary goals that the team is trying to achieve. They understand what this project means to the company's business plan or reputation.

Subcontractors, Vendors, and Suppliers

While they are not direct employees of the contractor, they are bound to the prime contractor by subcontract or purchase order. Subcontractors, vendors, and suppliers are critical assets to the team. They perform the lion's share of the work on almost every project. They maintain their own agenda yet are required to follow the lead of the prime contractor in the execution of the work. A major subcontractor can have a continual presence on the project from beginning to end and may mean the difference between success and failure of a project. They play a key role in both time and cost. They should be selected based on value and a right fit for the project, not just price.

The Design Professional's Group

The design professional's group is responsible for the design of the project. They are charged with translating the owner's requirements into tangible deliverables—contract documents, while complying with the local codes having jurisdiction over the project. There are numerous disciplines that can contribute to the project's design. This group's ultimate goal is to produce a comprehensive set of contract documents with plans and specifications that can be bid and then constructed in accordance with codes that have jurisdiction over the project.

Architect

The architect's team is the lead in the design group. They are responsible for the production of the plans that graphically represent the owner's requirements and the specifications that qualitatively define how the work will be performed. The architect is the *core* design professional responsible for the coordination between the various design disciplines (mechanical, electrical, civil, and structural engineers, etc.). This is no small task considering the technologically complex projects that are commonplace today. They are responsible for designing the project in compliance with all of the various building codes. The architect traditionally orchestrates the bidding process and aids the owner in evaluating and selecting the contractor. As the interpreter of the documents, the architect has a tremendous responsibility to the project team after the project has been awarded to the contractor. The most commonly used contracts provide for the architect to be the administrator of the contract during the construction process. This role is called *construction administration*. The construction administrator (CA) is not always the same architect that does the design. These are often very different skill sets. The construction administrator is actively involved in the execution of the work. Construction administration includes ensuring that the work progresses within the context of the documents, resolving discrepancies with the documents, clarifying or answering questions that arise as the work progresses, reviewing and ratifying change orders, approving submittals, and certifying applications for payment. Their contribution to the team commences with the design and ends when the project reaches Final Completion. Their responsibilities on the team are significant and the timely performance of their duties can contribute to a project's success.

They are key contributors to project meetings and represent the owner in many day-to-day decisions on the project. They are an ally to the contractor.

Engineering and Specialty Consultants

Due the highly specialized nature of the design process, the architect relies on engineers from a variety of disciplines to supply the specialized expertise required to meet the owner's needs. The engineer is responsible for designing the system that will fit within the architectural or *core* documents. The engineers include structural, civil, mechanical, and electrical to name but a few of the more common disciplines. Specialists in the specific use of the structure are also used such as technology wiring experts, kitchen designers for restaurants, or even acoustic engineers. These subconsultants function in much the same way as the architect during the construction process. They are called in to answer questions or clarify an aspect of the design. As the specialists that are responsible for many of the essential building systems, their duties to the team are numerous and invaluable.

The engineering consultants often speak the same language as the subcontractor performing the work they designed. Good relationships between the parties often make the process more efficient and smoother.

The Owner's Group

Not all owners are construction savvy. For many owners, the expansion of a facility or the construction of a new office is necessary for them to conduct or grow their business. The process itself, however, represents many unknowns. For many, it is not a regular occurrence; for others, it is a cyclical process: grow, expand, grow, expand. In all cases, the owner requires a firm or individual that can impartially represent them when required and make decisions to maintain progress. That function is traditionally performed by an owner's representative such as the construction manager.

Owner's Representative

The owner's representative traditionally is familiar with the business of the owner. They can be a single individual or a construction management firm. As a result, they are in a unique position to resolve issues relating to the needs of the owner in terms of project requirements. They can be full-time employees or consultants. They provide approvals on change orders in a timely manner and ensure that the funds flow so as not to delay the project. A well-qualified owner's rep has extensive design and/or construction background and direct and unfettered access to the owner. On many projects, the owner's rep is the eyes and ears of the owner and may be on-site full-time. They have decision-making authority. The owners rep's duties involve ensuring that the owner provides timely responses and information when required. Their contribution to the team is to ensure that the owner does not delay the contractor and to guarantee that the owner gets what he or she pays for.

Clerk of the Works

For large or complex projects, the owner may also have a second individual to oversee the work. Since most contracts between the owner and architect do not require that the architect be on-site full time, the owner (or architect) employs a *Clerk of the Works* to observe the progress and technical aspects of the construction. The clerk's main duty is to ensure that the structure is built in accordance with the plans and specifications. He or she records events, observes the work, and reports back to the owner or architect. They traditionally have little or no decision-making authority but are major contributors of the information essential for timely and accurate decisions. Their responsibility to the team is to ensure that observations and records are accurate, and the information is provided to the decision makers in a timely manner.

Construction Manager

For many larger, more complex projects the owner will engage a construction management firm to represent them in the design and construction of the project. CMs, as they are known, are brought on early. They oversee the initial budgets and schedules and are influential in the hiring of the design team. There are a variety of contract models for CMs, but all basically allow CMs significant authority in decision making for the project. This is seen as a layer of management between the owner and contractor or architect.

The Regulatory Group

The regulatory group, while not under the direct control of the project manager, is still a crucial member of the team. They have the ultimate ability to affect progress. They are traditionally a governmental agency (or agencies) with legal jurisdiction over the project. While securing approvals for zoning, site plan, and conservation should be finalized in the design development, well in advance of construction, this is not always the case. Team members should check to see that all required permits and "sign-offs" have been applied for and received. A stop work order issued by a local agency can have a detrimental effect on progress, not to mention the cooperation of the issuing agency. In most cases a cooperative and cordial relationship is the goal with this group.

Inspectional Services

Inspectional services are comprised of the authority that issues the permit as well as the entity that ensures compliance with the building code the permit is issued under. Depending on the locale and the project itself, there can be numerous permits that must be secured: building, plumbing, gas, electrical, fire alarm, HVAC, fire sprinkler, and street opening, to name but a few. Review periods and the issuing of a permit can be time-consuming and should be reflected in the project schedule.

As part of the Pre-Construction Planning phase, it is important for the PM and super to know what inspections are needed and any unique permits or approvals required both prior to commencement and at varying stages of the work. There is nothing that can slow or stop progress like a missed inspection that requires dismantling completed work.

It should be noted that often milestones in the project schedule center around a successful inspection such as rough electrical or plumbing, insulation inspection, or the testing of a major system such as the fire alarm.

The Plan—the Roadmap

Up to this point, we have emphatically stated the value of planning in advance of starting the physical work. We have identified the key personnel and their responsibilities. We have discussed how planning forces detailed thinking about the actual construction process. We have reinforced that policies and procedures are essential for an effective and efficient system to manage the work. However, we have been silent on what the deliverable or end result is of the pre-construction planning process.

The deliverable of the planning process is the *plan*. The plan is the executable roadmap as to how the project will be constructed. It is the project team's agreed-upon way in which the construction will satisfy the contract requirements. It is arguably the single most important by-product of the project team's work and will be the basis of the schedule and all future decision making.

All construction projects, due to their relatively short life span and intense use of resources during that life span, require a formal detailed plan to get the work done within the schedule and budget. It just doesn't happen as some miracle of nature. Hours of brainstorming, analyzing, and considering of possibilities produce the plan.

Despite the fact that there are some physical relationships that cannot change in construction (rebar must be set before the concrete is placed), there is still considerable leeway in determining how to accomplish many of the other tasks. Even if the entire project was a series of physical relationships with zero flexibility, there is still the need for a plan, due to all the differing agendas among the parties in the process. The plan must take into account these needs and agendas (to varying degrees) of all the parties. The plan is the systematic approach to the project: flexible enough to accept changes as they occur yet disciplined enough to allow for controls necessary to measure its success. The plan must have buy-in or a consensus that it is the best approach to the construction. Without a plan, we have no way in which to measure progress. Without a plan, what are we comparing the progress against? How can we hold people and parties accountable when there is no basis for measurement?

The plan should not be confused with the schedule. The schedule is the plan set against time and occurs later in the planning phase. The plan must define the work in sufficient detail so as to set the expectations of the participants, and in return, to ensure that they know what is expected of them. The plan must reduce the uncertainty or risk inherent in specific tasks. This is done by evaluating the potential risk and exploring ways to mitigate or eliminate it. The plan should establish the expected production or efficiencies of the participants, and in doing so it creates the basis for monitoring and controlling the work.

The plan is a series of decisions made with a focus on the future—the deliverables and goals of the project. It is a methodical organizing of the resources to achieve those goals within the fixed constraints of budget and schedule.

Each team member may do planning on their own and within their sphere of influence and for their own personnel and resources. Certainly, we can understand how the electrical contractor might create a plan to do specific tasks within his or her subcontract, but something that may reach beyond their own sphere of influence, such as energizing the permanent power, may have consequences beyond the electrical contractor and merits discussion with the other team members.

Remember, the plan has a direct relationship to the estimate. It is not, however, a fatal flaw to deviate from the estimate in considering other options. For example, if the project was estimated to install the windows from the inside and no scaffolding was carried in the estimate for the exterior, then it is highly unlikely that the plan will call for scaffolding at the exterior for the window installation. That having been said, the planning process might evaluate other options and consider erecting scaffolding at the exterior as the most cost-effective and efficient way to install the windows. This is especially true if multiple trades could take advantage of the scaffolding at the same time and share the cost. The project team might deduce that by erecting scaffolding, time could be saved and that the project overhead costs saved in shortening the critical path will be the business case for following that course of action.

Even for relatively small projects, the plan may mature through several generations with refinements right up to the last minute. This should be expected. "One-and-you're-done" efforts rarely, if ever, uncover the risk or problems that may be part of the plan, and never garner the support needed to make them successful. It is brainstorming while still retaining a common-sense, practical approach to executing the work that is the best path to take. It is the brainstorming that forces detailed thinking about the project and often clarifies or even invalidates earlier assumptions about the work.

The planning must address alternatives to solutions of potential problems. Identify the potential problem(s) and propose solutions, multiple solutions if possible. The planners should prioritize the solutions and choose the best solution as the one to be included in the plan. Most problems in construction have occurred before, and experienced project managers have had to come up with workable solutions in the past. Many of the problem-solving skills are learned and experienced-based.

As you will see in the next section, once the plan has been accepted among the team, it can be set against time, and now emerges as the schedule. The schedule, along with the estimate, is key to integrated project control and, ultimately, success.

Establishing a Baseline for Schedule Control

Once all team members have signed on to the plan, the next step is to commit the plan to time. More specifically, the plan has to be broken down into individual tasks that are more easily managed from both a monitoring and measuring perspective. This section will be a discussion of how the schedule fits into the overall controls process. A comprehensive review of scheduling and how it is applied to project control will be covered in Chapter 4 The Schedule.

Arguably, the single most important tool at the disposal of the project manager is the project schedule. Planned correctly and implemented as a management tool, it can provide a continuous feedback loop for the project manager to use in the decision-making process. There are several different types of scheduling methods used in construction today. However, the most effective as a management tool is the *Critical Path Method* (CPM) schedule. In short, the theory behind the CPM schedule is that there are a series of sequential tasks that are linked from the start to the end of the project that, if delayed, will delay the completion date of the project. It is based on the fact that these tasks, called *critical tasks*, are linked or interdependent. The start and finish of each critical task impact the start and finish of the succeeding critical task.

The schedule is the decomposition of all of the tasks on the project into recognizable finite activities, tasks with a clear beginning and end. Consider the task of installing light fixtures in classroom 101. While we may have no idea as to the size of classroom 101 or the number of light fixtures in it, the definition of the task is far less ambiguous than a task such as electrical lighting.

For all of the tasks that comprise the schedule, there are characteristics that define the work. Each task has a time, or *duration*, measured in workdays, which it will need to complete. The duration of a task is tied to its amount or quantity and the productivity of the individual or crew that will produce the work. Both the quantity and the productivity are derived from the estimate. Each task in the schedule has a relationship to the task(s) before and after it (with the exception of the first and last tasks in the schedule). Each task has a value in dollars (or labor-hours) that can be established, again using the estimate as the source. Each of these tasks can be assigned to an accountable party or resource. Lastly, tasks are classified into three types: administrative, procurement, and production.

The schedule, once agreed upon by the team as the best representation of time and resources to be expended, now becomes the schedule that will be used to manage the project. This initial schedule that will be shared or *published* with all team members and other stakeholders is called the *baseline schedule*. As the name implies, this initial schedule becomes the basis for comparison of the work that is scheduled to take place.

The baseline schedule is also referred to as the "as-planned," "initial," or "target" schedule. It will be used to compare as-planned progress to actual progress. Baseline schedules should be in place prior to the commencement of physical work. There is only one baseline schedule. As things inevitably change, a modified version of the baseline called an *updated schedule* is produced. Updated schedules are produced at regular intervals, most commonly monthly. The updated schedule will show how changes in the project to date have impacted the baseline.

Tasks that start late (or early) will show the new start date as compared with the baseline. If no inputs occur from the project management team, all things remaining the same, the task will finish late. If it's a critical path task, all succeeding tasks will be delayed. The updated schedule allows the PM team to discuss options to regain the lost time. This is where the corrective actions occur.

The baseline schedule has a date on which the controls clock starts ticking. It is called the *data date* or *status date*. It is the date on which we start tracking the schedule, performance, and costs that will be used in the monitoring portion of the controls process.

The baseline schedule, while a major part of the controls process, is not the only basis for measuring what we need. It is only half the tools in the toolbox. On the cost side, we need a way in which to measure cost performance similar to schedule performance.

Establishing a Baseline for Cost Control

In the same manner in which we set a baseline for schedule control, there needs to be a baseline set for measuring cost performance. Remember, accurate project control is the integration of schedule and cost performance data. The baseline for measuring cost performance is called the *budget*. The budget is a direct derivative of the estimate that was the basis of the contract. The estimate is broken down into individual tasks or subsets of tasks as in the case of the WBS.

The initial budget is called the *original budget*. It will be used to compare the as-planned spending (costs) to the actual spending (costs). Just as the schedule is updated, the budget can and most likely will be updated. As change orders occur, the budget will change to reflect the modification resulting from the change order. The new updated budget is called the *current budget*. It is the original budget plus approved change orders to date. It has a data or status date just as the baseline schedule does. In fact, the dates should coincide. When a change order is added, it changes the original budget by adding (or deducting) cost elements from the budget. A change order can be a WBS in and of itself with its own CBS. In other words, the PM team can track the costs within the change order separately to determine if the actual cost was more, less, or the same as the planned cost. It should be noted that not all change orders will have a corresponding change (addition or reduction) in time. Some change orders have no change to schedule because they are not on the critical path of the schedule. That being said, the schedule should be updated as the same time the change order is added. This will be discussed in greater detail in Chapter 4 The Schedule.

As previously mentioned, the budget is a direct derivative of the estimate. However, it is not the estimate exactly. In addition to normal costs for material, labor, equipment, subcontractors and overhead, the estimate contains profit. While the exact methodology for creating a budget may differ from contractor to contractor, contractors do not view profit as a cost. Most budgets on the job cost side exclude the profit as part of the cost. The budget is silent on the original estimated profit.

The budget is equally as important as the schedule. In some corporate cultures, more weight and attention may be focused on the budget than on a timely delivery. The opposite may be true as well. A "get it done at all costs" attitude can be adopted. Projects with hard and fast delivery deadlines may become less relevant if the date is missed. Consider an improvement to a stadium for Superbowl that does not get done in time. Its significance is diminished on the Monday after Superbowl.

Costs associated with a task are tracked in *cost accounts*. They are the costs associated with the lowest level of detail in the *Cost Breakdown Structure* (CBS). The CBS is the budgeted cost applied to the WBS detailed cost accounts. *Cost elements*, which are a subset of the cost account, segregate the costs in the accounts to reflect material, labor, equipment, subcontract, and even labor-hour totals for the task.

Ideally, for each task in the schedule, there will be a cost associated with the performance of that task. Tracking both the cost and the time to perform the task is essential for defining the accurate status of that task and the project as a whole.

A more thorough discussion of the budget and the definition of costs will be covered in Chapter 5 The Budget.

The Communication Plan

As noted earlier in this chapter, communication between team members and other stakeholders is essential for project success. The plan maps out the flow of communications to team members and is an essential part of the overall plan. The communication plan outlines what, when, and how information will be shared among team members so that they have the most up-to-date information on which to base a decision.

It establishes a communication chain, or an order to the stream of information. While it is fairly common for the electrical contractor to speak with the electrical engineer because they speak the same "language," any information that results from that conversation still must flow through official channels to become part of the project record.

Updates of routine information that can be relied on and appear at regular intervals can be a time-saver. Being able to log on to a dedicated project website to see what the change orders approved or the RFIs that have been answered with the knowledge that the website is updated at the same time daily is part of a functional communication plan. Whereas having to send multiple emails or leave repeated voicemails about change order or RFI status can be a waste of valuable management time.

The goal behind any professionally developed management plan is to communicate the information people need to do their job. It furthers the goal of project completion in a timely and cost-efficient manner.

A communication plan limits the access certain members of the team may have to certain information. For example, it may be crucial for the project manager to see job cost reports; however, no general contractor would want such proprietary information to be distributed to the subcontractors or owners.

A good communication plan identifies:

- the individual groups that make up the team, and occasionally parties outside the active team that need specific information
- the information needs of that group or individual team members

- where the information originates from (its source)
- where the information that will be distributed is stored or will be posted
- the frequency of regular or routine communications
- how the information will be distributed: electronically via email, updated on a website, shared in databases or on ftp sites, in person at meetings, etc.
- who will be responsible for distributing the information
- who will receive the information

Although this may seem fairly self-explanatory, it is amazing how often people and information are omitted. While the omission of information may not be an intentional act, it has the same impact as if it were. A good plan for communicating project information cannot be overstated. The advantage of establishing a communication plan is that instead of responding to information requests, the flow of information is controlled, timed, and routine. It minimizes disruptions and creates a level of confidence and professionalism in the overall process.

Summary and Key Points

The importance of pre-construction planning and setting both budget and schedule for baseline controls cannot be over-emphasized. Pre-construction planning involves setting in place policies and procedures so that each action is effective and efficient. This leaves more time for the problem solving that is essential to great project management.

Means and methods for the plan are created and ratified by the project team and go on to become the detailed CPM schedule or baseline for measuring performance from a time reference.

Equally important is the original budget, derived from the estimate that will be the baseline for cost control.

Some team members are direct employees of the contractor; others are bound to the project through their contract with the owner. Still other team members are peripheral, such as the regulatory agencies that have jurisdiction over a project. Identifying the team and their individual responsibilities is essential to setting up project control and getting the project off to a good start.

When all of the team members function effectively and efficiently, and understand the communication plan, information flows freely and in a timely manner among the group, and the project progresses. This synergy of individuals working cooperatively to further the project goals becomes the project team. Each team member contributes to the information stream for the project and allows others to make decisions based on that information.

Key points of this chapter are:

- The key to initiating the project controls is to prepare a plan that can be agreed upon.
- The plan is the means and methods as to how the contract will be completed.
- The plan can then be set to time so that the project has a schedule, with Notice to Proceed and Substantial Completion dates.
- The schedule serves as the duration of the project as a whole and by default as the duration of the individual tasks that comprise the project.
- The budget is derived from the estimate, although for tracking purposes the costs may be organized differently into more manageable CBSs that are defined by the agreed-upon WBSs of the plan.
- For any plan to work effectively, it must be supported by a reliable communication plan, that has routine updates at specific intervals.

Chapter 3 Pre-Construction Planning: Questions for Review

1. Successful construction companies have policies and procedures for managing a project. Policies and procedures are designed to control an action in order to provide a predictable outcome. True or False?
2. The pre-construction phase is intended to evaluate the constructability of the project and the documents that will be used as the guidance system for that process. True or False?
3. Define the responsibilities of the project manager.
4. Define the term "means and methods."
5. Many of the costs associated with a construction project are time-sensitive. Define what is meant by time-sensitive in this context and provide some examples.
6. Explain the difference between the plan and the schedule.
7. What is the baseline for time management and control on a project? Explain.
8. The initial schedule that is published for use as a management tool is called?
9. The project estimate is broken down into costs for measuring fiscal performance. This breakdown process is called?
10. Name three key components of a good communication plan.

CHAPTER 4

THE SCHEDULE

In previous chapters the reader was introduced to the concept and background of project management and the idea that the schedule and the estimate are different but very much interrelated. Up to this point in the text, the planning discussion has centered on identifying the work to be performed to achieve the deliverable. We have defined how each team member from the various groups contributes to the plan. With the plan flushed out, our efforts now turn to time management: the time to perform each task and the overall time to perform the project. Before one can understand the complexity of the relationship between budget and schedule, a review of the basic concept of scheduling, and the CPM schedule in particular, is necessary.

This chapter reviews scheduling in general and the Critical Path Method (CPM) of scheduling in particular. It is not meant to be a detailed explanation of CPM, but merely a refresher, focusing on how the schedule is used in project control.

Basic Scheduling Concepts

Scheduling is a decision-making process that is logically complex but mathematically quite simple. It is a step-by-step guide through the project with necessary time to perform each of the steps, or activities. The simple part is that schedules identify tasks and the time it takes to perform them. The complexity arises when those tasks are put in a sequence that is impacted by the performance of preceding tasks. Construction projects are essentially a series of tasks performed in a specific order for physical or organizational reasons. These tasks are built one upon the next, and the delay in performing one task may delay the start of its successor task. This is the essence of interdependence.

Construction schedules have some basic similarities, the most common of which is that work is measured as a function of time. That metric is the *workday*. Another is that the project is decomposed into incremental parts called *tasks* or *activities* that represent an action

Project Control: Integrating Cost and Schedule in Construction, Second Edition. Wayne J. Del Pico.

required to achieve the deliverable. For example: "Place and finish concrete slab at Area A." These tasks are performed in a specific order and often are interdependent. Conversely, some tasks are independent. Tasks can occur sequentially or simultaneously. There is also more complicated sequencing of tasks. The sequencing of tasks is referred to as the relationship between tasks.

Tasks in a schedule must be adequately described so that a non-involved party can understand what is occurring in the task. For example, the appropriate description is "Place and finish concrete slab at Area A" in contrast to just "concrete slab." In general, tasks consist of an action, a location, and a quantity. While the quantity is not required in the description, it is necessary in calculating the duration in workdays. Tasks also consume resources and money in addition to time.

A schedule must have milestones or interim goals for measuring incremental progress toward the final goal of completion. Tasks should also have a resource assigned to perform each task. The resource can be an individual tradesperson or a crew of multiple trades. Progress is contingent on the crew size and will change as the size of the crew is changed.

Ideally, though not always the case, resources should be used efficiently in the schedule. Tasks consume time to perform, and so each task must have an adequate duration to accomplish the task. The duration of the task is measured in workdays, although the overall schedule is measured in calendar days. Workdays are established as Monday to Friday and a single workday is eight hours (plus lunch and breaks). Major holidays are typically non-workdays. In contrast, there are seven calendar days in a week, and the week in the Gregorian calendar is measured from Sunday to Saturday. Workdays can be changed to include Saturdays and Sundays to reflect the specific needs of the project schedule. Most schedules are generated, at least at the start, based on a Monday to Friday workweek with the weekend as free time.

Schedules must be logical; that is, they must make sense both technically as well as chronologically. They usually follow the sequence dictated by normal construction practice, where physical relationships come first, followed by contractual relationships, and then managerial relationships. A physical relationship is one that cannot be changed. For example: the concrete foundation is cast and *then* the structural steel is set on it, not the other way around. Contractual relationships might include the phasing of a project, or specific deliverables required under the agreement. A managerial relationship is re-sequencing tasks to make a task more cost-effective or time-efficient.

The execution of the work assumes that each workday will produce the same amount of work. While most professionals understand this to be more of accepted practice than truth, a schedule assumes that over the duration of the task there is an average daily productivity that is the result of dividing the quantity by the daily output of the crew.

In addition to the text portion which describes the task, its duration and the start and finish dates of the task in schedules employ a very powerful visual tool that defines the work with symbols and illustrates the relationships between tasks.

Each of the following sections will explore the relationship between scheduling and project control.

Types of Schedules

There are numerous types of schedules in the construction industry, some of which are complex, others very simple. Regardless of the type of schedule used, it is the main vehicle by which the project team conveys it plan to satisfy the contract requirements of the rest of the stakeholders. It tells the participants where they should be, what they should be doing, when, and for how long. This allows the participants to plan their work in advance by selecting the appropriate crew to achieve the performance required within the time allotted.

Here are some of the common types of schedules that one would encounter on an average construction project.

The Checklist

While the checklist is not considered a formal type of scheduling, it is certainly prevalent at all levels of the industry and merits a brief discussion. Sometimes called a "to do list," the checklist is used for very short-term planning, rarely more than a few days. As the name would imply, it is a list of steps in a process to be performed in order to achieve an outcome. Its main purpose is to ensure that nothing is omitted. A checklist is an internal, rudimentary form of scheduling rarely shared beyond the author. It is highly effective in forcing the detailing thinking that is the essence of the schedule. Checklists have a simple tracking function: Check the work off as it has been completed—simple, but highly functional! They do not require special skills, equipment, or training to use.

Checklists are not always designed as a function of time. Due to their brevity, the tasks on the checklist are expected to be complete in the present or very immediate future. For longer periods the checklist can be combined with a simple calendar. Many software applications have the ability to print a Monday to Friday calendar with tasks assigned each day. Most handheld smart phones or tablets have the same capacity without the use of paper. For checklists that are time-sensitive, an alarm can be set on the device as a reminder.

More sophisticated checklists can be developed using an electronic spreadsheet such as Microsoft Excel™. Tasks can be listed in a column

Table 4.1 Checklist Using Microsoft Excel™

Ref. no.	Task	Duration	Due Date	Resource responsible	Milestone
1	Grade and compact trench bottom	2 days	7/5/2023	Site Contractor	No
2	Inspect bottom of footing	1 day	7/6/2023	Building Inspector	Yes
3	Verify footing locations, pin corners	1 day	7/7/2023	Civil Engineer	No
4	Deliver rebar	1 day	7/7/2023	Rebar Supplier	No
5	Form footings	5 days	7/12/2023	Formwork Contractor	No
6	Set rebar at footings	3 days	7/13/2023	Formwork Contractor	No
7	Cast concrete in footings	1 day	7/14/2023	Formwork Contractor	No
8	Strip footings	1 day	7/17/2023	Formwork Contractor	No
9	Erect one side of wall formwork	2 days	7/20/2023	Formwork Contractor	No
10	Set rebar in walls	2 days	7/21/2023	Formwork Contractor	No
11	Install sleeves in foundation wall	1 day	7/24/2023	Electrical Contractor	No
12	Double up panels and brace walls	3 days	7/25/2023	Formwork Contractor	No
13	Verify wall panel locations	1 day	7/26/2023	Civil Engineer	No
14	Cast concrete in wall forms	1 day	7/28/2023	Formwork Contractor	No
15	Strip wall forms	2 days	7/31/2023	Formwork Contractor	Yes
16	Break ties and patch holes	2 days	8/3/2023	Formwork Contractor	No

format with completion dates, milestones, and the responsible party. Table 4.1 is an example of a checklist made using Microsoft Excel™.

Disadvantages of the checklist relate to its simplicity. There is no way to show the interdependence of activities. If one task is delayed, what is its effect on the remaining tasks in the schedule? While extremely helpful, checklists are not really schedules and have limited applicability in that capacity. They are, however, extremely effective tools for completing tasks.

The Schedule Board

For many construction companies, a *schedule board* or *operations board* is a useful tool. This is a list of tasks or operations by project with individual employees or crews assigned by days or weeks. Calendar versions can be used as well with days of the week in a row across the top and tasks in a column along the left side. Crews or individuals performing the work are assigned by writing their name in the corresponding intersection of date and task with an erasable marker. Again, this is not a pure form of scheduling, but it does let those who perform the work know where they need to be, when, and for how long.

Disadvantages are again related to its simplicity. Rescheduling of tasks or resources can be time-consuming, especially rescheduling crews. Also, the schedule board is centrally located within a shop or assembly room. Personal copies are not typically distributed.

Checklists and schedule boards have their place and function. They should not be overlooked due to their simplicity. At the same time, they are not meant to be used as interactive scheduling method and have no system for tracking performance. Both are internal scheduling devices that are rarely shared beyond the author or company.

The Bar Chart

One of the most common types of schedule in the construction industry is the *bar chart*. The bar chart is also called the *Gantt chart* after its founder, Henry L. Gantt (see Chapter 1 The Basics). The bar chart uses a graphic representation of the project plan. It is traditionally developed with a chronological listing of tasks along the left side in columnar format. Along the top row is time in days, weeks, or months. Tasks are shown in their entirety, and each task is a highlighted bar from the start to the finish of the task. Figure 4.1 is a portion of a bar chart schedule.

The main advantage of the bar chart is its visual properties. It is easily understood and is a highly successful "big picture" tool. It is frequently used in presentations to senior management, the client, or the general public, when the detail of a CPM is unnecessary or too complicated. It is not meant to manage the work as it lacks the interdependency between tasks that is required for tracking, identifying trends, and forecasting outcomes. The bar chart does not allow the project manager to visualize the

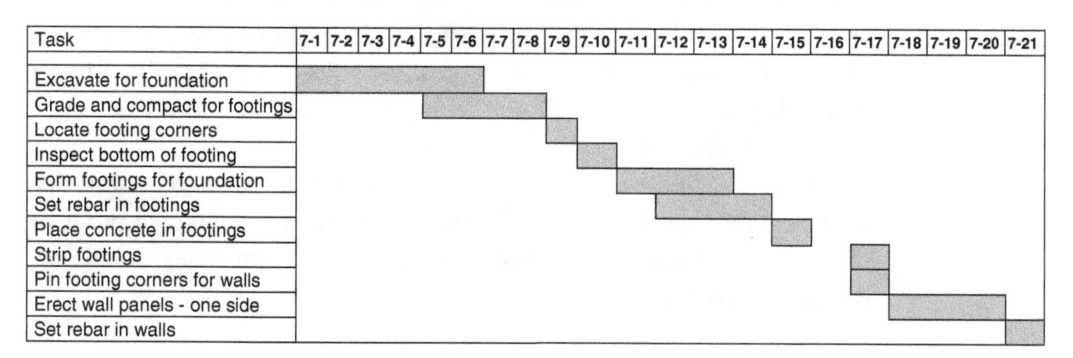

Figure 4.1 Bar Chart Schedule

project in specific terms. The bar chart requires updating by hand and is devoid of the logic included in scheduling software. Bar charts may identify the resource responsible for performing the task but rarely the cost associated with the task. Its greatest disadvantage is that it does not differentiate the tasks that are critical to achieving the project goals from the noncritical or concurrent tasks. It does not show how the delay of one task will impact the performance of another task. Its advantages are limited.

The Look-Ahead

Another type of bar chart is called the *Look-Ahead*. The Look-Ahead is a type of micro-schedule that provides a snapshot of a short period of time in the schedule. The most common time periods are two-week and three-week durations. As the title suggests, these micro-schedules focus on the next 10 or 15 workdays in great detail. They allow the team to analyze upcoming tasks or milestones to ensure that progress is met. Look-Ahead schedules are not typically used alone, but in conjunction with a CPM schedule. They are never meant to take the place of a real schedule, as they don't show the impact to the Substantial Completion date. For all intents and purposes, they are schedule checklists.

The Linear Schedule

Not all projects are suited to bar graphs or CPM schedules. So far, all of the schedules we have discussed are based on the premise that they can be broken down into separate activities, which can be analyzed and sequenced to find the best performance or activities. The problem with certain projects is they lack the distinctive segments of work to fit the CPM mold.

Consider the construction of an airport runway. Despite the size of a runway, it is essentially the same at either end or anywhere along the way. Therefore, the decomposition into segments of work defined by resource, material, or location commonly employed in CPM does not work.

Another type of scheduling, which takes advantage of the linear features of some projects like our example, is called *Line-of-Balance scheduling* or *velocity diagram*. These types of scheduling belong to the family of linear scheduling methods and will not be the subject of this chapter beyond a cursory introduction.

The Line-of-Balance schedule employs a simple visual chart that replicates the way linear work is done. It plots the cumulative progress of the work graphically against time. Figure 4.2 illustrates a Line-of-Balance schedule.

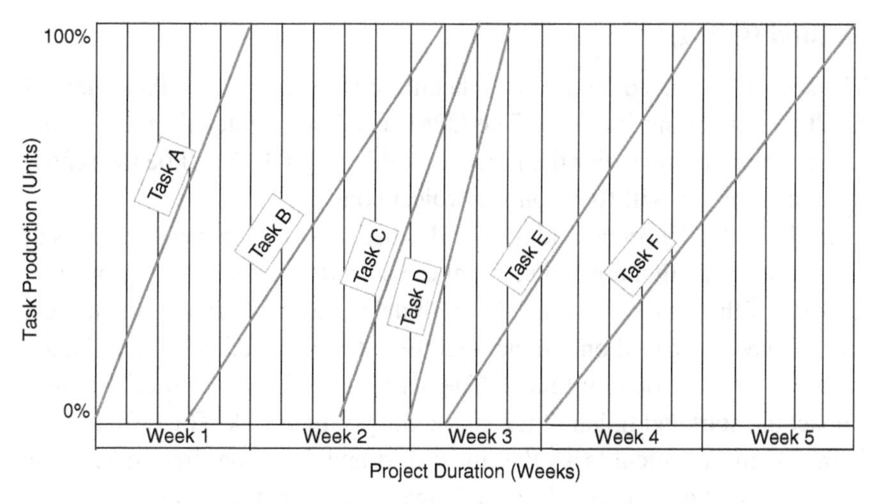

Figure 4.2 Line-of-Balance Schedule

Advanced Scheduling Methods

Our discussion of scheduling techniques has centered on basic methods; the checklist, schedule board, and bar chart. More sophisticated methods exist and are used regularly with great success in the construction industry. The two most well-known are *Program Review and Evaluation Technique* (PERT) and *Critical Path Method* (CPM). Both are uniquely different but used in specific project applications in construction.

Program Evaluation and Review Technique

Program Review and Evaluation Technique (PERT) has had its greatest acceptance in research and development. This is largely due to the probabilistic nature of research and development projects. However, PERT is not altogether foreign to construction projects. In fact, even though construction as an industry has an enormous experience base for most tasks, making it relatively simple to calculate durations, there is still a high degree of uncertainty on large, complex, and one-of-a-kind projects. PERT assumes a high degree of variability in task durations, making an accurate single estimate of duration more difficult to arrive at. PERT uses three estimates of duration: optimistic, most likely, and pessimistic. A formula is then used to calculate the appropriate duration of the task based on rules of probability.

While the PERT method of scheduling may not be appropriate for most construction projects, there is definitely some applicability to calculating the duration of high-risk, critical path tasks. PERT can be used to analyze the probability of a specific range of durations for a task, thereby allowing the scheduler the opportunity to accept or reduce risk. PERT calculations will be addressed in detail in Chapter 13 Risk Management.

Critical Path Method (CPM)

The second advanced method of scheduling is called Critical Path Method or CPM scheduling for short. The CPM schedule is uniquely fitted to the construction industry. For the purpose of this text, CPM will be the scheduling method we will focus on for project control.

In short, the theory behind the CPM schedule is that there are a series of sequential tasks that are linked from the start to the end of the project. If any one of these tasks is delayed, it will delay the finish date of the project as a whole. It is based on the fact that these tasks, called *critical tasks*, are separate but inextricably linked. The start and finish of each critical task impact the start and finish of the succeeding critical task. The *critical path* is the chain of critical tasks that must start and finish on time to keep the project as a whole on time. It is the longest path of planned activities to the finish of the project with the shortest duration.

Not all tasks are critical; some tasks can occur at the same time as critical tasks and are called *concurrent* tasks. Critical tasks can also be defined as those tasks in which the amount of time available (in the schedule) to perform a task is equal to the time required to perform the task. For concurrent or non-critical tasks, the time available is always greater than the time required to perform the task. This extra time is called *float* and its duration is measured in workdays. There are several varieties of float and each will be discussed later in this chapter.

By its very design, the CPM allows the project manager to compare actual progress to planned progress. The CPM also employs some of the graphic format of the bar chart although with the added component of interdependency. Interdependency results from a predecessor task impacting the successor task. Even non-critical tasks have predecessors and successors. It is the linking of tasks in such a way that it represents the planned work in an efficient and logical manner. The network logic or the relationships between tasks represents the majority of the work of creating the schedule. Even with a computer this can be a time-consuming process that requires a lot of thought.

While the overwhelming majority of practitioners schedule with some type of software, it is essential to understand what goes on inside the CPM scheduling process. This text assumes the reader understands the mechanics behind CPM and will avoid a discussion of the forward and backward pass and the arithmetical computations associated with them.

The CPM Schedule as a Management Tool

Since its inception in the 1950s, CPM has proven to be the most effective scheduling method for displaying and tracking project information necessary for managing and controlling time on a construction project.

The single most important tool at the disposal of the project manager is the project schedule. Planned correctly and implemented as a management tool it can provide a continuous loop of feedback for the project manager to use in the decision-making process. As previously noted, the theory behind the CPM schedule is that there are a series of tasks that are linked and, if delayed, will delay the finish date of the project. It is this concept of linked tasks or *interdependency* that allows the project team to create a working schedule that can act as a *baseline*. The baseline is the series of values—start, finish, and duration—that can be used for comparison to what actually happens when the work is performed.

When the initial schedule is published, the participants get to review what is expected of them and when. By reviewing the start date and duration, they can determine when materials and/or equipment need to be delivered as well as the size of the crew that will be needed to perform the work. It also allows the manager to know what work has to be complete prior to the start of this work, and what other work is contingent upon this work being complete. The schedule allows the manager to compare the progress of the work as a whole, not just a specific task. As the tracking portion of the schedule is updated, the project manager can compare actual progress to the planned progress and spot trends, both positive and negative.

The actual progress of the project or a task tracked against planned progress can be used to determine which subcontractor or crew is ahead, on schedule, and more importantly, which ones are behind schedule. It can be used to manage subcontractors as well. It is very easy to demonstrate using CPM that at the continued daily output, the critical task will not be completed by the planned finish date and the task will be late. This will impact downstream tasks on the critical path.

For tasks that are ahead of schedule and appear like they will finish early, the updated schedule can be used to move up the succeeding task to exploit the lead. If the PM does not track the performance, they may know things are proceeding better than planned, but will not know when to start the succeeding task. The same is true in reverse. A task that is going long and will go beyond its finish date, can be used to resequence the succeeding task if possible.

The Baseline Schedule

As discussed previously, the initial schedule that has been developed as a result of the agreed-upon means and methods (the plan) to accomplish the work is called the *baseline schedule*. Done correctly, this is the fruit of many hours of creating and resequencing a list of tasks, calculating durations, and assigning relationships that become the network logic. It includes adding deadlines and constraints and any other reasonable tool for managing the project. Preferably, this has been reviewed and accepted by major subcontractors, and in some cases is the direct result of contributions from those subs and vendors that have a continual presence on the project. The in-house team: superintendent, project staff, and senior management have contributed to this schedule, and it has their buy-in. In some cases, the owner and the architect have weighed in on the schedule and approved it, or at the very least reviewed it. (Owners and architects very rarely *approve* a schedule due to the legal implications. They most often *review* it and acknowledge that it complies with the contractual obligation of providing a baseline schedule.)

The last step is to publish or share the schedule with those who will be stakeholders in the project. In an ideal world the PM team would set the baseline with the software and upload it to a hosting website for the project. It would be followed up with a notification to those who have access to the site, that the schedule has been posted and is now active.

When the baseline is set in scheduling software, it freezes the start and finish dates of the project. Of equal importance is that it freezes the start and finish of individual tasks, especially those on the critical path. As the work progresses and the team starts tracking performance with schedule updates, they compare the actual start date, actual duration, and actual finish date with those set in the baseline schedule. Because dates in the baseline are fixed, it makes it easy to determine if a task started late (or early), ran long, or finished earlier (or later) than planned in the baseline. The team can then determine the cause of the delay, assuming there is one, and determine a corrective measure to bring the schedule back in line with the baseline.

Baseline schedules have the distinct advantage of illustrating, with great clarity, the time impact a change order has on the critical path. This makes for a challenging argument to deny an extension of time is due for the change order.

One final thought about the baseline schedule. The PM team should remember that the schedule is part of the contractor's means and methods that they contribute to the project. The schedule is the prime or general contractors's domain. They create it, manage by it, and live and die by it. Projects often have penalties for late delivery: liquidated or actual damages. As a result, resequencing tasks or denying legitimate extensions of time to accommodate an owner's or architect's image of what the schedule *could* be, is a foolhardy move. Evaluate all suggestions to improve the schedule but bear in mind the final decision is that of the project manager. Remember the schedule is created by those who will be responsible for it. It is always better to shorten the schedule by increased performance than to have to request additional time because you capitulated to another party's request.

Tasks

Tasks are the activities that make up the schedule. They consist of an action that consumes time, requires labor-hours from resources, may involve equipment, and almost always costs money. Tasks have a beginning and an end, which are referred to as start and finish dates. Tasks move the project forward toward completion. Determining which tasks should be included in a schedule and which are not required can be a challenging assignment. Identifying individual tasks or combining subordinate tasks to arrive a composite task is also difficult. There is one source of information that is the basis of the tasks that will comprise the schedule. It is the estimate. The estimate details how the estimator planned to construct the project, and, though it may not be a step-by-step guide, it contains the information that the professional can use to create the schedule. The tasks in the estimate are not imported verbatim into the schedule. We schedule differently than we estimate. For example: the estimate for a slab-on-grade may be a series of tasks: place vapor barrier, set welded wire mesh, place concrete, finish concrete surface, and cure concrete slab. This may be a single composite task in a schedule, composed of the other subordinate activities. It might appear as *Place and finish slab-on-grade* in the schedule. There are some loosely defined rules for what tasks end up in the schedule and which don't. This may require that we identify what we want to track and group it with other subordinate tasks. This is done by analyzing the estimate.

In the past, the term *task* was used to describe an action or deliverable in an estimate. The very definition of the word task implies work being done. The term *activity* was reserved for an action or deliverable in a schedule. Similarly, the term activity envisions action and getting things done. This distinction between terms has all but disappeared. The terms task or activity with regards to an actionable item in a schedule will be used interchangeably here.

The Work Breakdown Structure

This process of analyzing and deriving the tasks from the estimate is called *deconstructing* the estimate. The same deconstruction process will be used in Chapter 5 The Budget, to develop the budget that will serve as the baseline for cost control. The deconstruction process begins with the estimate that represents the contract amount, sometimes referred to as the *as-bid* estimate. The estimate will list all of the work encompassed by the contract, so it is a good place to start in determining a list of tasks for the schedule.

The estimate is reviewed and then broken down into a series of Work Breakdown Structures (WBS) (see Chapter 3 Pre-Construction Planning). The WBS is a model that divides the project into logical or physical sections. These sections often follow the organizational system of the estimate called CSI MasterFormat™ (see Chapter 5 The Budget). The WBS is composed of multiple layers or levels that start with a big picture or summary view and work their way down to the details. It is not uncommon for a WBS to have three to five levels of breakdown, with the detailed task being the lowest level. This following is an example of WBS for scheduling:

- Level 1: Division 9–Finishes
- Level 2: Section 09 30 00–Tiling
- Level 3: Sub-section 09 30 16–Quarry Tiling
- Level 4: Tasks for Schedule
  - Prep subfloor for quarry tile flooring.
  - Install quarry tile and cove base.
  - Grout and clean quarry tile.

The level of detail required in the schedule is predicated on the level of control needed to manage the project. Clearly, for some projects, the three tasks previously outlined could be summarized in a single task. In most cases that would be more than sufficient especially if it was performed by a subcontractor.

While each project manager may have their own special way to develop a task list, there are a few general considerations that may be helpful:

- Too much detail can be intimidating and can render the schedule useless as a management tool. Participants will ignore it.
- Remember it is not a set of instructions. The trades performing the work know what is entailed to reach the deliverable.
- With too little detail, tasks are omitted. Vague schedules cannot be used as a management tool.
- More detail is required at extremely critical stages or sequences.
- The schedule should have milestones to measure interim progress as key points in the schedule.
- Deadlines for owner/stakeholder decisions that impact the critical path should be identified.
- Tasks that lead to a milestone should be detailed so nothing is omitted that might prevent the milestone from occurring.
- Tasks that are performed by owners, inspectors, and design team members should also be listed.
- The task should result in a deliverable that can be measured.

It is recommended that the PM team start with more tasks than may be required and combine or shed tasks that do not add value to the tracking of the work.

Task Types

Every project manager understands that there is far more to a schedule than just the tasks of installing the physical components of the work. There are permits to be issued, submittals to be approved, equipment to be ordered, and items to be fabricated off-site, to name but a few. All of these items have a tremendous impact on the production work, and without them, there would be no production tasks.

Consider the erection of structural steel on a building project. Before the steel can be erected it has to be fabricated, and before the fabrication, there has to be detailed shop drawings drafted that show each piece and how they are connected. These drawings are reviewed and approved by the structural engineer, or corrections may be needed and a resubmittal may be required. In any event, it is evident that each step in the process is necessary and contingent upon the step or steps before it. Imagine if the shop drawing review took four weeks longer than scheduled. Would that delay the fabrication of the pieces? Without a doubt. If the fabrication is delayed, then one can understand how the delivery would be delayed, and then the erection of the steel delayed, and so forth all the way to the end of the project.

So, if a task such as shop drawing review and approval can have that dramatic an impact on the schedule and the actual production work, should it be included in the schedule? The answer is a resounding yes! The action of shop drawing review is outside of the control of the contractor. If this action can derail the progress of the steel fabrication and erection, it should be noted in the schedule so that it is conveyed to the design team the importance of the review time to the schedule.

A brief note of caution is due here. With the previous scenario, one could make an easy argument that each and every action from the signing of the contract to the submission of the last closeout document can have an impact on the schedule and, therefore, should be included as a task in the actual schedule. Wrong! This is where professional experience and judgment reign supreme. If too many tasks, especially unnecessary tasks, are included in the schedule, the schedule becomes impossible to manage and track. Worst of all, the schedule can no longer function as a management tool; it becomes too large even for a small project. It has become a set of instructions to build the project instead of a guideline. Just as too little detail renders a schedule too ambiguous to be of use, too much detail can intimidate, confuse, and result in the schedule being ignored. Even though there are administrative and procurement tasks for every item on the project, not all need to be included in the schedule.

There are a few considerations for deciding if a task should be included in the schedule, but the reasoning can vary from project to project. The ultimate decision belongs to the project manager and the project team.

- Is the task on the critical path?
- Does another entity not in the project manager's direct control have responsibility for this task?
- Will this task require a resequencing of succeeding tasks to keep the schedule on track if it is delayed?

An answer of yes to any of the above may make the task worthy of a place in the schedule, but not always. It needs to be analyzed on a case-by-case basis. The superintendent may also be consulted for what he or she views as a schedule-worthy task. As noted earlier, not every task is a production task. Tasks or activities can be categorized into one of three general types:

Administrative activities are support tasks that are subordinate to the actual construction process. They are crucial to the contract side but not directly part of the physical work itself. They include tasks such as submission of structural steel shop drawings or review and approval of steel shop drawings.

Procurement activities are the actions required to ensure materials are delivered to the site for installation. They include tasks such as ordering roof top units or fabricating and finishing interior casework.

Production activities are the elements of physical work that comprise the project. These activities have a direct relationship to construction. They require materials, crews, and equipment. They include tasks such as setting rebar at the foundation wall or erecting joist and deck at the second floor. These tasks are used to measure progress.

Task Descriptions

While there is a no universal convention for describing tasks in a schedule, the following should be considered when describing an activity:

- Describe the task in terms of an action being taken, such as place concrete, frame deck, apply waterproofing, etc.
- Describe the task in terms of the building component involved such as footing, columns, door, slab, etc.
- Describe the task in terms of where the action is taking place or the geographic location of the building component such as basement footing, second floor panel, southwest wall, etc.
- Describe the task in terms of timing, or stage, such as phase 2 demolition, rough-in electrical power, etc.

- Describe the task in terms of organizational or contractual responsibility, such as rough inspection by building commissioner, sign-off by owner's representative, or architect review and approval of shop drawings, etc.

Some other general points to consider in preparing a list of tasks:

- Remember that we estimate differently than we schedule. A task in an estimate may be priced in five separate lines of the estimate, whereas we might schedule the task in one line.
- When deciding which administrative tasks should be included, a good rule is to review the technical specifications sections, especially Part 1 General. This part describes required submittals, approvals, handling, coordination, and compliance.
- Review the Division 1 General Requirements section of the specifications. It is a good source for contractual requirements, such as phasing.
- Start with a large, detailed list of tasks, and then consider what items can be removed or combined. This is often easier than trying to increase the list of tasks.
- Study the estimate carefully not only for tasks, but also for means and methods. This can generate tasks or add to the description of a task.
- Separate tasks by the entities that will perform the work. It is not good to combine multiple trade or subcontractors in an activity, especially if payment is based on progress.

Most schedulers create a list that follows the sequence in which the project would be built, the "ground up" concept. When the scheduler is stuck, he or she can always ask, "What comes next?" This is helpful in creating and refining the logic of the schedule. Nobody wants to look at a schedule that shows the taping of the drywall *before* its hung.

The difference between the task description *Excavation* and *Excavation for footing along Column Line 1* is as stark as night and day. Remember the schedule communicates the plan of the work.

Task Durations

Duration is defined as the time it takes to execute a task. It is the difference between a task's start date and its finish date. Durations are measured in days, more specifically workdays. As noted previously a workday is 8 hours and typically Monday to Friday. Measuring durations in hours would be a tracking nightmare from a project control perspective. Not to mention that there are hours in the workday in which crews are stocking materials, setting up for work, reading plans, giving, or getting instructions, having a break, or several other less than productive, but necessary tasks.

These hours would skew the productivity. Using weeks as the increment would not reflect the fact that some tasks take only a single workday to complete. The week as a frame of reference would be too long.

Workdays are also the appropriate increment because there is a direct correlation to productivity and the daily output of the individual or the crew. Productivity is measured by the daily output for a crew. A superintendent can "measure" the work done for the day by what work has been done since the crew started and ended for the shift. The topic of productivity and its relationship to project control will be discussed in detail in Chapter 10 Productivity.

Determining the duration of a task is less a matter of experience than a mathematical calculation. That is not to say that experience does not play a part. There are two primary methods in which individual task durations are determined. Both require that a few general rules are followed:

1. Assume that each task will be performed normally. In this circumstance, the term normal defines ideal or near ideal working conditions: temperate climate, appropriate tools, adequate materials close by, sufficient working space, etc. This ideal set of conditions translates to the most efficient productivity for the crew and, as a result, the lowest unit cost. It is not unrealistic to expect that a task will be performed at the most efficient productivity. In fact, it is fairly common to have a task proceed unimpeded. Assuming normal productivities allows room for schedule compression if it should become necessary.
2. Evaluate each task independently of predecessors or successors. Assume no other work existed. While professionals understand this is not realistic, it portrays the work as unimpeded by other work. The impacts of other tasks or constraints will be calculated when the schedule tasks are linked.
3. Use consistent units to measure the duration; the accepted unit, as noted previously, is workdays. This prevents confusion and the requirement to convert other units to measure time.
4. Document how durations were determined, what productivity rates were used, and why the rates were modified if they were. If a productivity rate was diminished, document the logic behind the reduction. This allows the project manager to defend the calculation later if the duration comes into question. This is especially helpful in situations where a condition changed from what could have reasonably been anticipated during the bid process.
5. Assign an assumed crew based on realistic productivities and the space available to do the work. Note that for the application of most productivity logic, it is accepted that if 1 carpenter will trim

8 windows in a day, 2 carpenters will trim 16 windows in a day. Again, this assumes sufficient room in which to work, enough windows to trim, sufficient tools and materials, etc. This logic does not hold true in all cases. At some point, productivity will diminish due to a variety of reasons. Finding the sweet spot between crew size and productivity may require multiple attempts.

Calculating Durations

As mentioned before, there are two basic methods for calculating task durations. Both methods involve productivity and a reference back to the estimate. Remember, the estimate holds the key component required for determining how long a task will take to execute. That component is its quantity. Reasonable logic suggests that the larger the quantity of the task, the longer it will take in workdays to perform unless the crew is increased.

Daily Production Rate Method It is fairly common for companies to track the work they self-perform. Tracking work performance has two distinct advantages: It validates the unit price carried in the estimate; and it allows the scheduler to determine from actual performance records how long it will take to perform that task. This is especially true when the crew has limited experience on performing the task.

Published cost data such as that produced by the RS Means Company provide productivity rates for a wide variety of tasks and crews. These tasks, their relative productivities, and crews can be compared to individual projects as another check and balance.

The most common and simplest method of calculating task duration is called the *daily production rate method*. It is based on the productivity of a specific crew and the assumption that the productivity of this crew will remain constant over the life of the task, provided that the conditions under which the work is performed do not change. This method recognizes that as the crew size changes, the daily output will change. The daily production rate method is based on the simple formula:

$$D = Q \div DO$$

where:

D = the total duration of the task measured in workdays

Q = the total quantity of the task being performed by the crew in units of measure for the task

DO = the daily output of the task performed by the crew-day in units of measure for the task per day

It should be noted that the units of measure for both Q and DO must be the same. Consider the following example:

Assume a crew of two carpenters has a daily output for hanging drywall on metal stud partitions of 2,000 square feet (SF) per day. The quantity takeoff portion of the estimate has the total square footage of drywall to be hung on metal stud partitions at 13,800.

$$\text{Duration} = 13{,}800\ \text{SF} \div 2{,}000\ \text{SF/day} = 6.9\ \text{days}$$

The daily production rate method is somewhat less flexible than the second method especially when the crew size is varied significantly. In the daily production rate, productivity has a direct relationship to crew size.

Labor-Hour Productivity Method The second method focuses less on the crew size than the time it takes to install a single unit of the task. Most estimators use cost per unit of measure ($/SF, $/LF, $/SY, etc.) to estimate a task. There is a direct correlation between the cost ($) and the labor-hours to perform the task. Hence, the cost can be converted to the amount of time in labor-hours regardless of the crew size. Many regard this method to be more flexible and even more accurate for scheduling. It can even be used to determine the size of the crew needed to perform the task in the allotted time.

The labor-hour productivity method uses the following formula:

$$D = TLH \div LH/d$$

where:

D = the total duration of the task measured in workdays

TLH = the total quantity of labor-hours of the task being performed derived from the estimate

LH/d = the total labor-hours worked per day on the task

Consider the following example:

Historical data reveals that a single square foot of drywall takes 0.008 labor-hours for a carpenter to hang on metal stud partitions. The total of drywall to be hung on metal stud partitions from the estimate is 13,800 SF. 13,800 SF × 0.008 labor-hours/SF = 110.4 labor-hours.

$$\text{Duration} = 110.4\ \text{labor} - \text{hours} \div 16\ \text{labor-hours / day} = 6.9\ \text{days}$$

While both reveal that same duration in days, the labor-hour method allows the flexibility to adjust the crew size as needed to fit the available duration for the task. Again, within reason.

A note about experience and where it fits in is appropriate. In roughly the same manner as the estimator adjusts costs based on actual conditions, the scheduler uses his or her prior experience to judge how close the actual task is to *ideal*. For example, if a task is determined by the scheduler to be more difficult to perform in real life than it would under ideal conditions, the scheduler may want to extend the duration of the task to account for the reduced productivity. Schedulers adjust the calculated durations to reflect the complexity and unique conditions of each individual project. Other considerations would include external influences, such as the weather, or even providing for a learning curve. Sometimes the adjusting is no more than the rounding up of partial days to full days. Remember that calculating durations too conservatively is planning for failure. Other more complex adjustments to duration to mitigate risk will be discussed in Chapter 13 Risk Management.

Calculating task durations with historical data from self-performed work can be relatively straightforward. However, a large portion of the work on every general contract is performed by subcontractors. This can present a different set of challenges, especially when the scheduler has limited expertise in the subcontractor's field. Whenever possible, the subcontractor should be consulted and encouraged to provide durations by one of the previously mentioned methods. For subcontracts that have not been let out as of the time the schedule is developed, the scheduler should consult published data or other team members with more experience in the matter at hand. Guessing is strongly discouraged. In addition, unrealistic productivity, or crew sizes just to fit the work in the time available, is pure fiction. Those performing that scope of work will ignore the schedule as unprofessional and unachievable.

Interdependencies

The CPM schedule has a particular feature that mimics the real-life construction process. It is the *interdependency*, or relationship between tasks. Consider the process of constructing a cast-in-place concrete foundation wall. After erecting a single side of the wall formwork, the horizontal and vertical rebar within the wall has to be set, then the second side of the panel can be erected. Both sides of the formwork panels have to be erected and braced before the concrete can be placed inside the forms. These are physical relationships that cannot be altered or ignored. They are the basis of the construction process and are expressed in the logic diagram or *network* that is the essence of the critical path method of scheduling. This diagram is the means of expressing the sequence of operations as well as the anticipated outcome when the task is complete.

The graphic side of the schedule illustrates how each activity's performance will affect those activities downstream from it. It takes very little experience to acknowledge that if one of the steps is delayed, the steps that follow will be delayed, almost in a chain reaction fashion. It is this interdependency between tasks in the schedule that has endeared CPM to the construction industry.

When developing the logic diagram that will show how tasks depend on other tasks, the scheduler must ask three simple questions:

1. What other task(s) must be completed before this task can begin?
2. What task(s) cannot begin until this task is completed?
3. What task(s) can proceed at the same time as this task, without interfering with this task's execution?

The answers to these three questions will help the scheduler define one task's relationship to another.

Many tasks follow a standard relationship. The simplest of all task relationships is the *finish-to-start* relationship. It is the condition in which the first task or *predecessor* task must be complete before the following task or *succeeding* task can start. It is the most common relationship found between tasks on a schedule. As an example, consider that a doorframe must be installed *before* the door can be hung on the frame. Interdependency means the task of hanging the door must wait till the installation of the doorframe is complete. If the frame installation is delayed, it logically follows that the hanging of the door will also be delayed. This is due to the physical relationship between the two tasks that no amount of creative scheduling can alter.

Frequently, we do not have to wait till a task is complete to start the next task. In fact, some tasks can start together and run concurrently. This type of relationship is called the *start-to-start* relationship, which synchronizes the start dates of two or more tasks so that they proceed simultaneously. (This should not be confused with two tasks that are totally unrelated and have no relationship.) As an example, consider the installation of insulation within a partition as the drywall is simultaneously being hung. Just prior to each sheet of drywall being hung, the insulation is installed within the stud bay. This has its best success when both tasks have approximately the same duration, although it is not a requirement.

Last in the category of simple relationships is the *finish-to-finish* relationship. As the name might suggest, the finish dates of the tasks are linked to end at the same time. The finish dates are the driver. This is sometimes used when a milestone is contingent upon the finish of multiple tasks or when all work must be completed to start a new task or phase. Consider the scenario in which the installation of the doorframes and the tape and finish of the drywall must finish on the same date so that painting can start the next day.

Complex Relationships—Lags and Leads

Unfortunately, not all relationships are quite as simple as the doorframe and the door. To make the best use of time and resources, subcontractors are scheduled to follow behind one another in a wave and compress the waiting or down time. It is similar in concept to the landing of jet airliners at busy airports. Landings are separated by miles or minutes. As one jet touches down and taxies off the active runway, another is on final approach, and yet a third is maybe three miles out lining up for a landing.

In the construction industry, this separation between the subcontractor's start or finish dates on the schedule is called a *lag* or a *lead* depending on the relationships between the tasks.

Consider the start-to-start relationship defined in the previous section, but instead of starting both tasks simultaneously, delay the start of the second task to allow the first task to progress. Imagine that task A is the framing of a metal stud partition, and task B is the rough electrical wiring within the partitions for power. To economize the time, task B does not have to wait until task A is complete. It is logical that instead of starting at the same time, the start of task B should be delayed a period of time from the start of task A to allow the carpenters to install an adequate quantity of partitions so that the electrical contractor's crew has sufficient work ahead of them to maintain an efficient workflow and productivity. The relationship between task A and task B would then be categorized as a *lag* relationship based on its delay in start dates. In practice, if we delayed the start of task B for two days to allow the carpenter to erect two days' worth of partitions prior to the start of the electrical rough wiring, it would be stated that task B has a start-to-start relationship with a two-day lag. Figure 4.3 illustrates a start-to-start relationship with a lag.

When the relationship is based on the finish dates, as noted in the finish-to-finish relationship, and there is a separation of time, it is referred to as a *lead* relationship. For example, if it were essential that the partition framing be completed two days in advance of the completion of the electrical

Day-1	Day-2	Day-3	Day-4	Day-5	Day-6	Day-7	Day-8	Day-9
		Task A						
					Task B			

2-day lag

Figure 4.3 Start-to-Start Relationship with a Two-Day Lag

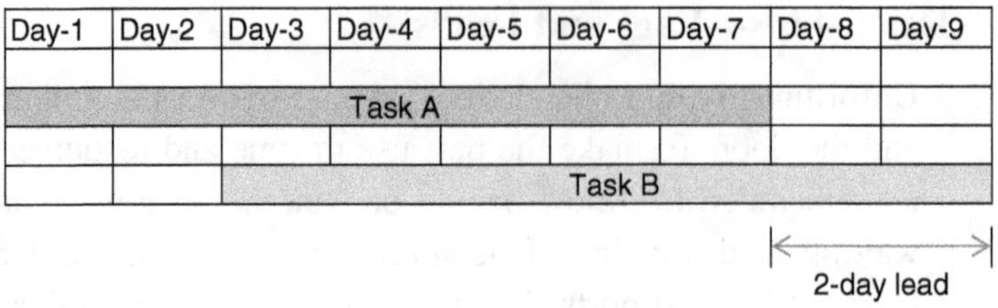

Day-1	Day-2	Day-3	Day-4	Day-5	Day-6	Day-7	Day-8	Day-9
			Task A					
					Task B			

2-day lead

Figure 4.4 Finish-to-Finish Relationship with a Two-Day Lead

rough-in, it would be stated that task A has a finish-to-finish relationship with task B with a two-day lead. Figure 4.4 illustrates a finish-to-finish relationship with a lead.

It is frequently confusing as to when it is appropriate to use a lag and when to use a lead relationship. The general rules are:

- Start-to-start lags are used when the preceding activity is *faster* that the succeeding activity.
- Finish-to-finish leads are used when the preceding activity is *slower* than the succeeding activity.

For both complex and simple relationships, it is the interdependency resulting from the links that will govern when a task starts or finishes relative to another task. It is this concept that provides one of the baselines for project control.

For the novice scheduler, a good practice is to avoid attempts to complicate the schedule with complex relationships. These can often present a challenge in determining what caused changes to overall project durations. Remember the software only performs the calculations as directed by the relationships we create. Once the schedule has been built, the scheduler can go back and try "what if" changes to see their impact.

Milestones and Constraints

One of the crucial steps in preparing the schedule is to establish intermediate goals along the critical path of the schedule. These intermediate goals are called *milestones*. Milestones act as way markers to measure the progress of the project as incremental portions of the work get accomplished. They allow the project manager to compare actual progress to the anticipated progress on the baseline schedule. Milestones play a big part in the monitoring process within the feedback cycle. The unique feature of milestones is that they mark the time in which a specific task or group of tasks are to be complete. It is a barometer of both time and progress.

In order for the milestones to be an effective indicator of progress, milestones have to be selected carefully in accordance with a few basic guidelines. These guidelines include the following:

- Milestones should be on the critical path. It is the critical path as displayed in the baseline schedule that drives the progress of the project.
- Milestones should be distinct, natural markers that can be easily understood and identified by all parties. Abstract achievements that have little relevancy to the majority do not make good milestones.
- Milestones should be naturally occurring in the construction process. They should be goals for the trades to visualize and focus on.
- There should only be a few milestones in the schedule. Too many can dilute their significance.

Classic examples of milestones that have a tangible, practical value in indicating progress include:

- rough electrical or mechanical inspections
- rough frame inspection
- structure made tight to the weather
- completion of steel erection
- permanent power energized
- temporary certificate of occupancy

Milestones can be the completion of a phase with multiple tasks contributing to the achievement or can be the completion of a single task signifying a major turning point in the project.

Another type of marker on the schedule is called the *constraint*. A constraint is loosely defined as restriction on the start or finish dates of an activity or series of activities. Constraints have a large role in the project control process since they may have financial penalties associated with their dates. Constraints are used to ensure that a task, or tasks, meets a specific deadline. Constraints also mark when a task must start or finish, before or after a specific date.

There are a variety of constraint types used in scheduling a construction project, the most common of which are date constraints. Date constraints include *deadline* and a delivery date. A deadline is defined as a "no-later-than" constraint. Deadlines are attached to the start or finish of a task to indicate an absolute condition. Another type is called the "start-no-earlier-than" constraint that is applied in the case of a delivery date. For example, a delivery date for the delivery of structural steel requires that any succeeding linked task, such as the erection of structural steel, would have a start-no-earlier-than date to coincide with the delivery date.

Date constraints are typically added after the schedule has been created and should be applied sparingly. Some other types of constraints include:

- *Start constraints*: Stipulates that a task must start no-earlier-than or no-later-than a particular date. This is similar to a deadline.
- *Finish constraints*: Stipulates that a task must finish no-earlier-than or no-later-than a particular date.
- *Mandatory constraints*: Stipulates absolute dates that a task must-start-on or must-finish-on. Regardless of predecessor tasks. Mandatory constraints force the task to start or finish by a specific date.
- *Default constraints*: Stipulates that a task will start-as-soon-as-possible or start-as-late-as-possible. Start-as-soon-as-possible is frequently the default constraint that most software employs. It takes advantage of any shortened duration of a predecessor by moving the start date of the successor up.

Whatever the circumstance, constraints are powerful factors in the timing of tasks. Too many constraints reduce the flexibility of the schedule should the resequencing of events become necessary.

Float

Float is defined as the difference between the amount of time available to perform a task and the time required. Float is measured in workdays, and is reserved for tasks that are not on the critical path. By definition, tasks on the critical path have a float of zero.

There are several types of float that can occur in a schedule:

- *Total float* is the time that a task may be delayed without impacting the overall project completion date.
- *Shared float* is the float available across a string of noncritical tasks. It is sometimes called "string" float. If the first task in the string uses all of the shared float, there will be no float available for subsequent tasks in the string.
- *Free float* is the amount of time a task can be delayed without delaying the early start of successor tasks. If the task uses its free float, no other task will be affected.
- *Independent float* is the float that belongs to a single task and that task alone. Free float can become independent float if all the predecessors finish and there is float available.

Regardless of the type, float is the project manager's friend. It allows tasks to be delayed, resequenced, or rescheduled without delaying the project as whole. Float can allow a project manager to move resources to a more critical task without causing a delay.

No discussion of float would be complete without the age-old question that inevitably rears its head... "Who owns the float?" If a handful of stakeholders were asked the question, there would be a handful of responses, all of which are based on their individual position in the process. The majority of those responses would be incorrect. The answer is simple. The project owns the float. The correct question is... "Who controls the float?" To that question the answer is the general or prime contractor. As mentioned earlier in this chapter, the schedule is part of the means and methods the general or prime contractor brings to the project. They are responsible, contractually for the success of the project with regards to the schedule. It is their responsibility to dole out or restrict the shared or other types of float as they see benefit the project. At one point or another all parties to the contract are impacted by float or lack thereof. It is not up to one subcontractor to use all of the string float to the detriment of the other subcontractors in the string. The GC or prime assigns the float. The owner can often benefit from the float. Consider a project that finishes early, most owners will take possession of the project early. Even if they elect not to take possession willingly, the contract provides in most cases that Substantial Completion is the defining date when the owner becomes responsible for the project.

Updating the Schedule

Once the baseline schedule has been published and work is progressing, things change. The plan that was the basis of the schedule may need resequencing, tasks may have been delayed, and tasks may even have finished early. This is all part of the day-to-day events that mark an average construction project. However, since the schedule is meant to be a management tool, it needs to have the most up-to-date information. Without it, the decision making by the project team might be flawed. It is important to keep the schedule accurate by regular updates. Most contracts provide for routine updates of the schedule just prior to or at the submission of the Application for Payment. The idea is that payment is tied to progress. Even without a contractual requirement for updating the schedule, regular updates are essential for realistic project control.

The *cycle* or interval of the update is dependent upon the duration of the project. The shorter the duration, the more frequent the update cycle. Clearly, from a practical management and control perspective, a six-week project would need updates more often than once a month; however, monthly may be adequate for a 36-month project.

An update is intended to compare actual progress to planned progress as illustrated by the baseline. It can be as simple as establishing the actual start and finish dates of a task. It can be the reporting of work-in-progress as a percentage of completion, or it can be the resequencing of tasks due to changes. Delayed or accelerated work should also be reflected on updates, as well as change orders issued during that period. Each will have an impact on the schedule.

The information that forms the basis of the update is derived from the record keeping that is part of project control. Records such as the daily reports, payroll records, RFIs (Requests for Information), equipment hours expended, subcontractor personnel on-site, or even material delivery slips add to the data used in updates. Each can contribute to an accurate snapshot of the project status. All of this is combined to create the historical record of the project. Over the duration of the project, the regular updates compared with the baseline paint an accurate picture of the performance of the participants. The more frequent the update cycle, the less work involved since the amount of information to be analyzed is less.

As will be discussed later in the text, update information must be accurate, especially when determining the percentage of work that has been completed. Project managers should shy away from opinions as a basis of determining percent complete and instead rely on physical counting or measuring procedures. Opinions are subjective and can be the source of dispute between team members. Progress can be measured in both percent complete and days remaining. This is typically a preference based on the software used for reporting.

Updating the schedule provides the project team with answers to questions such as:

- Is the project progressing as scheduled?
- Is the plan that is the basis of the schedule the most efficient way in which to execute the work?
- Does work need to be resequenced?
- Are there tasks that were omitted from the baseline that are now necessary?
- Are the approved change orders impacting the schedule?
- Are RFIs being answered in a timely manner so as not to delay the work?
- Are subcontractors performing in sync with the schedule?
- Are the administrative and procurement activities in the schedule on track?

Updating the schedule is critical to evaluation project status and management decision making.

While project schedule updates may seem daunting, they are usually not. Remember for those tasks we subcontract, the updating of the project is big picture. We measure the performance of the task on the schedule by whatever metrics are available. We count light fixtures, or measure flooring installed, or calculate tons of steel erected. Those self-performed tasks may take more effort and analysis. The superintendent or foreman may keep running totals on the quantity of the task completed. The best part is the sheer quantity of tasks that require updates are less due to the interval of the update.

Summary and Key Points

A major part of the integrated control process is the project schedule. The schedule is the plan discussed in Chapter 3 Pre-Construction Planning, set to time. The project schedule, once complete and agreed upon, becomes the baseline for the progress and performance measurement from the time perspective. Performance is based on the relationship or interdependency between tasks on the schedule.

There are simple rules as to which tasks should appear on the schedule or which tasks are unnecessary. Too much detail and it can be intimidating, too little detail and it is useless as a management tool.

Tasks have durations and are based on crew sizes. They are not just arbitrary durations plugged in to fit the available time. Some tasks fall on the critical path of the schedule, and some occur concurrently. All tasks require monitoring at regular intervals. Regular updates to the schedule allow the team to analyze the status of the project and discover what, if any, decisions or corrective actions are required.

It is not just the schedule that is tracked but the integration with costs, as we will see in Chapter 5 Budget, that provide the clearest and most comprehensive picture of project status.

Key points of this chapter are:

- Schedules are intended to measure work as a function of time. The best type of schedule for construction is the CPM in that it reflects the interdependency between tasks on a construction project.
- The CPM schedule can be used as a management tool to compare actual progress to the baseline schedule.
- Tasks are the individual activities that occur in a schedule. Tasks originate from the estimate. They have quantities that are used to calculate duration.
- There are two methods for calculating the duration of a task. The Daily Output Method that focuses on the size of the crew and the Labor-Hour Method that determines overall hours.
- In addition to work tasks, schedules can have milestones and constraints that reflect progress or deadlines.
- Not all tasks in the schedule are production tasks. Some tasks are of an administrative or managerial nature such as shop drawings and decision deadlines.
- Non-critical tasks have float. The float belongs to the project but is controlled by the general or prime contractor.

Chapter 4 The Schedule: Questions for Review

1. The construction schedule measures work as a function of time. True or False?
2. The incremental parts or work of the schedule are called?
3. What are required to measure incremental progress toward completion?
4. A velocity diagram is used for scheduling what type of work?
5. What type of scheduling assumes a high degree of variability in task durations?
6. Define the basic concept of CPM scheduling. Cite the relationships between tasks and why it is pertinent in construction.
7. Define the term "float." Explain how it can benefit the project.
8. Tasks can be classified into three types. What are the three types?
9. Task durations are measured in workdays, overall schedules are measured in calendar days. Explain the difference.
10. Explain the difference between the Daily Production Rate method and the Labor-Hour Productivity method for calculating durations.

CHAPTER 5
THE BUDGET

Since the beginning of this text, a consistent theme has been reinforced: The schedule and the estimate are different but very much interrelated in project control. It is the integration of both schedule and cost that allows the project manager the accurate control needed to manage the work. Having reviewed the scheduling process in the project control paradigm, it is now time to consider the budget and its role in project control. In short, for every task in the schedule, there is a corresponding cost for the work. Since the task originated in the estimate it makes sense that the cost will be there as well. This chapter reviews the estimating process and its contribution to project control. The author assumes that the reader has a basic understanding of estimating and its role in the construction process.

Fundamentals of the Estimating Process

Of equal importance to how long the project will take is how much it will cost. All construction projects have costs associated with them and for the contractor, as well as the owner, it is essential that costs be known in advance. While it is impossible to know exactly what the project will cost in advance of the work, it is very possible and practical to estimate, or accurately approximate, the cost in advance.

Contractors, like most other businesses, have a strategy for performing work. A big part of that strategy is that the work is profitable, and the way work becomes profitable is to start by pricing it correctly, then managing the work efficiently, with the price as a yardstick for cost performance. Traditionally, contractors chase work that is in their area of demonstrated expertise. Loosely stated, everyone in the organization has a skill set that contributes to estimate, manage, supervise, or construct that type of project. Contractors develop a market niche by repeatedly doing similar types of projects: schools, roads, retail, office buildings, homes, etc. It is the repetition that allows the contractor's team to excel at the skills needed to hopefully make the project profitable. It is a learning process. Since a large majority of contracts are based on a stipulated sum delivery methodology, that will

Project Control: Integrating Cost and Schedule in Construction, Second Edition. Wayne J. Del Pico.

be the focus of our discussion in this section. A stipulated sum, often called a *lump sum* or *firm fixed price* delivery methodology, is based on providing construction services on a definitive scope of work for a fixed price.

Before beginning the discussion, a few definitions are in order. An *estimate* is defined as the anticipated, accurate, approximate cost of all materials, labor, subcontractors, equipment, and overhead associated with a particular construction project. Costs are anticipated because the estimate is prepared in advance of the work being performed. The adjective "accurate" speaks for itself; an estimate that is not prepared carefully with adherence to industry standards is misleading and worthless. Last of all, the costs are approximate as it is impossible to know what the exact cost is in advance of performing the work that will generate the cost.

Cost is defined as the price paid to acquire, produce, accomplish, or maintain something. So cost is relative to the frame of reference it is viewed from. For the owner, who will pay the general contractor (GC), cost includes the contractor's profit. For the GC, who will perform the work, cost is everything but the profit. However, the general contractor's cost, relative to a subcontractor's contract, includes the subcontractor's profit.

Since our perspective is that of the GC, cost will exclude the GC's profit, but will include the profit of the subcontractors and vendors that are carried in the GC's estimate. It should also be noted that the estimate is the in-house, proprietary work of the contractor, and is very rarely, if ever, shared with anyone outside the in-house project team. It could prove damaging if it fell into the hands of the competition. When the price and terms of the estimate are offered to the client, the estimate serves as the basis for the *bid*, or *proposal*. The bid (or proposal) is the contractor's offer to perform the work priced in the estimate in return for the compensation and terms defined in the bid.

An estimate is prepared using *bid documents*. Bid documents generally consist of plans, specifications, and *addenda* or changes to the bid documents or process. Plans are the graphic representation of the work. They show size, use of the space, and dimensions, and provide the information necessary to determine the quantities of the work. Plans quantify the project for the estimator. The specifications, or more aptly the *technical specifications*, define how the work will proceed. The technical specifications define acceptable practices, tolerances, handling, and related work. It also defines the products that will be included in the work. Specifications are qualitative in nature and set the acceptable standards that the quality of the project will be measured against.

An addendum (singular) is a bulletin that is issued that makes a change to the plans or specifications. Addenda (plural) are often used to answer questions or clarify something in the documents. Occasionally, they may be issued to change something in the bidding process such as bid date, time, or location. Used together, the plans and specifications plus any modifications from the addenda provide sufficient information to quantify and price the work that will be the basis of the contract.

The cost as defined above, from the GC's frame of reference, is all costs to complete the work: material, labor, equipment, subcontractors, and overhead, both direct and indirect. All of these costs in the estimate will be paid out to some entity involved in the work during the course of the project. The remaining funds between the contract amount and the costs paid out is the profit. Therefore, the contractor must track these costs to determine the profit.

Similar to scheduling, there are different types of estimates, but one has a natural affinity for project control.

Types of Estimates

The majority of competitive bidding in a stipulated sum contract is based on the *unit price* estimating method. Unit price estimating is the methodology by which bid documents are broken down into their incremental components for pricing. Done correctly, it is the most accurate way to approximate the costs of a project. It starts with the calculation of the quantity of all materials, labor, and equipment for each item of work, coincidentally called a task. This process is called the *quantity survey* or, more commonly, the *takeoff*. Quantities are then multiplied by the unit price, or the incremental cost of the work, and the result is called the *extended price*. This method is used by both general and subcontractors. Table 5.1 is an example of a unit price and its extended price.

Once the entire set of bid documents is taken off, priced, and extended, the totals for each category of work are summarized. Subcontractor proposals are substituted for the appropriate *plug* or placeholder. This compilation of estimated costs, called the *Estimate Summary*, or *Recapitulation* is marked up for indirect overhead and profit to arrive at the final number. This final number represents the estimator's vision of the cost of the work plus profit based on the selected means and method to execute the contract. It is based on the incremental costs for all of the tasks compiled to represent a single number.

Tasks in an estimate are similar to those of a schedule. They consume resources and time, and, as such, have a cost. Tasks in an estimate, exactly like a schedule task, must be adequately described so that a non-involved party can understand what is being priced in the task. For example, the appropriate description is "place and finish concrete slab at Area A" in contrast to simply "concrete slab."

Table 5.1 Example of Unit Price with Extension

Task description	Quantity	Unit	Unit Price	Extension total
Place and finish concrete slab	5,000	S.F.	$3.00	$15,000.00

Tasks in an estimate have quantities that define the amount of work to be performed in an individual line item or task. The quantity is one of the key features that link the estimate to the schedule. As stated in Chapter 4 The Schedule, the quantity is used to define the duration of a task based on the crew and daily output. In addition to the estimating value this adds weight to the correctness of the quantities. An error in quantity will not only provide an erroneous value for the cost, but will lead to an incorrect duration in the schedule.

Organization of Estimates

In contrast to scheduling, estimating does not necessarily follow an exact step-by-step, ground-up approach. Unit price estimating follows an industry-recognized organizational structure called CSI MasterFormat. CSI is the acronym for Construction Specifications Institute, the group responsible for CSI MasterFormat. It is the most widely accepted system for arranging construction specifications and estimates. The system is also used for classifying data and organizing manufacturers' literature for construction products and services.

CSI has allocated an eight-digit code and topic descriptions to all components of the specifications. MasterFormat groups the information in the project manual into four major categories:

- Bidding Requirements
- Contract Forms
- General Conditions
- Technical Specification sections

While all four categories are important to the estimating process, it is the last category, the Technical Specification sections, that is the focus for the discussion of project control.

MASTERFORMAT™ 2020, the latest version of MasterFormat consists of fifty construction divisions within the technical specifications section. It has been divided into five subgroups. Each division is a compilation of similar or related work numerically organized into subsections called "levels." Each level represents a further breakdown of the CSI division classification. With the publication of the 2004 edition, the MasterFormat numbers and titles were revised to allow them to cover construction industry subject matter more adequately and to provide ample space for the addition of new sections. The titles that make up MasterFormat were also revised, reflecting the new edition's renewed focus on work results. Since 2004, the organizational structure of MasterFormat is refined and released as a new year. Hence the latest version is MASTERFORMAT™ 2020.

As a part of this process, the numbering system was revised in its entirety. All section numbers and many section titles have changed from the 1995 edition. The five-digit numbers used in the 1995 edition were expanded to allow for more topics at each level of classification. The old numbers were limited at levels two through four to only nine subdivisions. Because of this limited number of available spaces at each level, many divisions of MasterFormat simply ran out of space to properly address topics. This lack of space often resulted in inconsistent classification. These limitations were solved by making the new MasterFormat numbers six digits in length and arranging the digits into three sets of paired numbers, one pair per level. These pairs of numbers allow for many more subdivisions at each level. Meanwhile, the main six-digit number still represents three levels of subordination, as the numbers in previous editions of MasterFormat have done.

For example, consider the CSI section number 03 30 53.40. The first group of two digits, 03, is MasterFormat Level 1 and designates the division the work belongs to, in this case Division 3–Concrete. The second group of digits, 30, is MasterFormat Level 2, which designates the subsection Cast-in-Place Concrete within Division 3. The third group of two digits, 53, is MasterFormat Level 3, a further breakdown of the Cast-in-Place subsection, Miscellaneous Cast-in-Place Concrete. The last group of two digits, .40, is MasterFormat Level 4 and deals with one component within the previous Level 3 section.

The fifty MasterFormat divisions were determined based on the relationships of activities in the actual construction process. They roughly follow the natural order of building construction. The specification divisions and a general summary of their contents are as follows:

- Division 00—Procurement and Contracting Requirements: advertisement for bids, invitation to bid, instruction to bidders, pre-bid meetings, bid forms, wage rates, bond forms, and related certifications.
- Division 1—General Requirements: a summary of the work; definitions and standards for the project and project coordination, meetings, schedules, reports, testing, samples, submittals, shop drawings, closeout, cleanup, quality control, and temporary facilities; pricing issues, such as unit prices, alternates, and allowances. This is often the source of the project (direct) overhead requirements.
- Division 2—Existing Conditions: existing conditions of the site or structure, demolition, surveys, geotechnical reports, salvage of materials, lead, asbestos and mold remediation, and the removal of underground storage tanks.
- Division 3—Concrete: formwork, reinforcing, precast, and cast-in-place concrete, concrete curing, and cementitious decks.

- Division 4—Masonry: brick, block, stone, mortar, anchors, reinforcement, and masonry restoration and cleaning.
- Division 5—Metals: structural steel, metal joists, metal decking, light-gauge framing, and ornamental and miscellaneous metals.
- Division 6—Wood, Plastics and Composites: rough and finish carpentry, millwork, casework, composite lumber products, and plastic fabrications.
- Division 7—Thermal and Moisture Protection: waterproofing, dampproofing, air and vapor barriers, insulation, roofing, siding, caulking, and sealants.
- Division 8—Openings: metal and wood doors and frames, windows, glass and glazing, skylights, mirrors, finish hardware, vents and louvers.
- Division 9—Finishes: gypsum wallboard systems, board and plaster systems, painting and wall coverings, flooring, carpeting, acoustical ceiling systems, and ceramic and quarry tile.
- Division 10—Specialties: items such as demountable partitions, toilet partitions and accessories, fire extinguishers, postal specialties, flagpoles, lockers, signage, and retractable partitions.
- Division 11—Equipment: specialized equipment for homes, as well as banks, gymnasiums, schools, churches, laboratories, prisons, libraries, hospitals, etc.
- Division 12—Furnishings: cabinetry, rugs, tables, seating, artwork, and window treatments.
- Division 13—Special Construction: greenhouses, swimming pools, integrated ceilings, incinerators, and sound vibration controls, and clean rooms.
- Division 14—Conveying Systems: elevators, lifts, dumbwaiters, escalators, cranes, and hoists.
- Divisions 15 through 20: Reserve divisions for future expansion.
- Division 21—Fire Suppression: fire suppression and protection systems.
- Division 22—Plumbing: plumbing piping; waste water, and vent, gas piping, special piping, refrigeration, and controls.
- Division 23—Heating, Ventilating and Air Conditioning: heating, air conditioning, ventilating, ductwork, controls, insulation, HVAC equipment, solar energy heating equipment, and humidity controls.
- Division 24—Reserve division for future expansion.
- Division 25—Integrated Automation: network servers, integrated automation of HVAC, fire protection, electrical systems, communications, energy management systems, etc.
- Division 26—Electrical: electrical service and distribution, wiring devices, fixtures, communications, and power.

- Division 27—Communications: communication services, cabling and cable trays for communication, adapters, and software.
- Division 28—Electronic Safety and Security: fire alarm systems, closed circuit TV, security alarm systems, access control, leak detection, and video surveillance.
- Division 29 and 30: Reserve divisions for future expansion.
- Division 31—Earthwork: clearing of the site, excavation, backfill and compaction, grading, soil treatments and stabilization, and heavy site work such as shoring, pile driving, and caissons.
- Division 32—Exterior Improvements: paving, curbing, base courses, unit paving, parking specialties, fences and gates, landscaping, and irrigation.
- Division 33—Utilities: piping for water, sewer, drainage and related structures, fuel distribution utilities, and electrical utilities.
- Division 34—Transportation: railways and track, cable transport, monorails, transport signaling and control, etc.
- Division 35—Waterway and Marine Construction: coastal and waterway construction, dams, marine signaling and dredging.
- Divisions 36 through 39: Reserve divisions for future expansion.
- Division 40—Process Integration: specialty gas and liquid process piping, chemical process piping, measurement, and control devices.
- Division 41—Material Processing and Handling Equipment: bulk materials handling and conveying equipment, feeders, lifting devices, dies and molds, and storage equipment.
- Division 42—Process Heating Cooling & Drying Equipment: industrial furnaces, process cooling, and drying equipment.
- Division 43—Process Gas and Liquid Handling, Purification, and Storage Equipment: liquid and gas handling and storage equipment, gas and liquid purification equipment.
- Division 44—Pollution Control Equipment: air, noise, odor, water pollution control, and solid waste collection and containment.
- Division 45—Industry-Specific Manufacturing Equipment: oil and gas extraction equipment, mining machinery, food and beverage manufacturing equipment, textile, plastic, and a variety of other types of manufacturing equipment.
- Division 46—Water and Wastewater Equipment: package water and wastewater treatment equipment.
- Division 47—Reserve division for future expansion.
- Division 48—Electrical Power Generation: fossil fuel, nuclear, hydroelectric, solar, wind, and geothermal electric power generation equipment.
- Divisions 49 and 50—Reserve division for future expansion.

Technical Specifications

The last of the four categories is called the Technical Specifications, or Technical Sections (Divisions 00-49), which define the scope, products, and execution of the work. These are the "meat and potatoes" sections that provide the estimator with the necessary information (in a highly organized and industry-accepted format) to accurately price and build the structure. The technical sections provide the following information for each activity:

- administrative requirements
- quality or governing industry standards
- products and accessories
- installation or application procedures
- workmanship requirements and acceptable tolerances

This information is organized within the section in three parts: general, products, and execution.

Part 1 General Part 1, the general section of the specifications, provides a summary of the work included within that particular section. It ties the technical section to the General Conditions and Supplementary General Conditions of the Contract, an essential feature in maintaining continuity between the general contractor and subcontractors. This is sometimes called the *flow-down* provision. Roughly translated, it allows the general contractor to assign responsibility for work to a subcontractor. Part 1 identifies the applicable agencies or organizations by which quality assurance will be measured. It defines the scope of work that will be governed by this technical section, including but not limited to items to be furnished by this section only, or furnished by others and installed under this section. It also identifies other technical sections that have potential coordination requirements with this section, and defines the required submittals or shop drawings for the scope of work described in this section. Part 1 also establishes critical procedures for the care, handling, and protection of work within this section, including such ambient conditions as temperature and humidity. If applicable, it addresses inspection or testing services required for this scope of work.

Part 2 Products Part 2 deals exclusively with the products and materials to be incorporated within this technical section of the work. For products that are directly purchased by the contractor from a manufacturer or supplier, the items can be identified using one of four methods:

- proprietary specification
- performance specification

- descriptive specification
- specification compliance number

Proprietary specification: Proprietary specifications spell out a product by name and model number. Proprietary specifications have the unique advantage of allowing the architect or owner to select a product they desire or have used successfully on prior projects. The advantage of requiring specific products is the level of reliability they provide. The disadvantage is that they eliminate open competition. To help reduce the exclusivity of the proprietary spec, the architect often adds phrasing called the "or equal" clause, which allows limited competition. While the "or equal" clause opens the door to some competition, it can be risky, as it puts the burden of equality on the proposing party (the contractor or the subcontractor, or even the vendor) who proposes the substitution. Proposed alternates should be reviewed closely. What may appear as a comparable product might not pass muster under closer scrutiny by the architect during review of the submittals or shop drawings. If the proposed substitution is not acceptable, the contractor is responsible for providing the specified product originally named in the specification, unless there has been some negotiated relief provided by the owner or design professional.

Performance specification: An alternate method of specifying products and materials is based less on makes and models and more on the ability to satisfy a design requirement or perform a specific function. This type of specifying is called a performance specification. In lieu of specifying a particular product by name, the architect opens competition to all products or materials that can perform the specific functions required to complete the design. This approach allows healthy competition among various manufacturers that have a similar line of products. It ensures competitive pricing and more aggressive delivery schedules.

Performance specs can identify products by characteristics, such as size, shape, color, durability, longevity, resistivity, and an entire host of other requirements. Some products that are not specified by name can be identified generically by reference to a particular American Society for Testing and Materials (ASTM) testing number or a Federal Specification number. Pricing materials or products by their conformance with an ASTM number presents risk as there could be several different grades of one product with vastly different prices. The architect makes the final decision as to whether a product has satisfied the performance criteria. The entity proposing the material or product should be able to prove performance compliance with comprehensive facts and evidence, such as copies of pertinent tests and their results, and manufacturers' data. For a specification section that involves custom-fabricated work, the language might be a mixture of proprietary and performance specifications.

Descriptive specification: The third method of specifying a product or process is by using descriptive specifications, which are written instructions or details for assembling various components to comprise a system or assembly. They are similar to a recipe. Most often, descriptive specifying is used for generic products, such as mortar or concrete. Frequently, no manufacturers or proprietary names are mentioned or needed.

Specification compliance number: Another method of product specification is by the use of a federal, ASTM, or other testing agency reference number. This method tends to be a favorite choice of the military and federal agencies. It holds the product to a very strict set of guidelines or tolerances established through testing and past performance. If the product contains the compliance number, it typically satisfies the requirement.

Part 3 Execution Part 3, called the Execution, deals exclusively with the method, techniques, and quality of the workmanship. This section makes clear the allowable tolerances of the workmanship. The term "tolerances" refers to adjectives such as plumb, straight, level, flat, or true. The Execution section describes any required preparation to the existing surfaces in order to accommodate the new work, as well as a particular technique or method for executing the work. Part 3 also addresses issues such as fine-tuning or adjustments to the work after initial installation, general cleanup of the debris generated, final cleaning, and protection of the work once it is in place. Some sections of Part 3 may identify any ancillary equipment or special tools required to perform the work, such as staging or scaffolding.

CSI MasterFormat technical specification sections have a direct correlation to the WBS and the cost accounts. They are an established, and near universally accepted organizational system that goes beyond the specifications to include the takeoff and estimate and even the data collection system and job cost reporting systems used in cost control. CSI is an organizational system that is pervasive through each of the steps of cost control, so it makes perfect sense to adopt it for project control. It can even be adopted with some minor challenges, to the scheduling portion of project control.

The Baseline Budget

For each of these divisions of the Technical Specifications and all of the items of work that comprise them, there is a cost. As noted in a previous section of this chapter, the recapitulation of these costs is called the Estimate Summary. If the contractor is the successful bidder and is awarded a contract for the project, the estimate and the Estimate Summary will become the baseline for cost control (Table 5.2). In other words, it will be used to measure the cost of a task by comparing the actual to the planned (or estimated) cost. It is the task and corresponding cost for the task that will represent a line in the budget.

Table 5.2 Estimate Summary Sheet Example

Bid Date:	**July 20, 2023**	**New and Renovated Office Building**				**BASE Bid**	
Time:	**2:00PM**	**Boston, MA**				**Addenda;**	**#1, #2**
Sect.	**Description**	**Materials**	**Labor**	**Equipment**	**Sub**	**Totals**	**Remarks**
01 00 00	Project Overhead Summary Sheet				$ 405,726	405,726	See Project Overhead Summary Sheet
02 41 00	Selective Demolition				52,200	52,200	Acme Demolition Company
02 81 00	Asbestos Abatement				6,950	6,950	ACM Abaters,Inc.
03 11 00	Cast-in Place Concrete (Formwork)			3,600	108,000	111,600	Cape Pumping Co./Smith Foundations, Inc.
03 21 00	Reinforcing Steel/WWF	17,557				17,557	Rusty's Steel Company, Inc. deliv. bar/WWF
03 31 00	Cast-in Place Concrete (Ready-mix)	26,554				26,554	Tri-County Ready-Mix Concrete
03 35 00	Cast-in Place Concrete (Flatwork)			7,169	26,350	33,519	Jones Concrete Finishing Co.
04 22 00	Concrete Unit Masonry				276,500	276,500	ABC Masonry Contractors
05 12 00	Structural Steel				156,200	156,200	State Iron Works, Inc
05 21 00	Steel Joists					–	State Iron Works, Inc
05 31 00	Metal Decking					–	State Iron Works, Inc
05 50 00	Metal Fabrications				80,735	80,735	Columbus Metal Fabricators
06 10 00	Rough Carpentry	7,973	13,248			21,221	Self-performed work
06 20 00	Finish Carpentry	16,785	53,486			70,271	Self-performed work
07 10 00	Dampproofing				17,600	17,600	All Weather Dampproofing, Inc.
07 21 00	Building Insulation	1,841			24,190	26,031	Self-performed work
07 53 00	Elastomeric Roofing/Flashings				126,780	126,780	Sky Roofers, Inc.

(Continued)

Table 5.2 (*Continued*)

Bid Date:	**July 20, 2023**	**New and Renovated Office Building**				**BASE Bid**	
Time:	**2:00PM**	**Boston, MA**				**Addenda;**	**#1, #2**
Sect.	**Description**	**Materials**	**Labor**	**Equipment**	**Sub**	**Totals**	**Remarks**
08 11 00	Metal Doors and Frames	31,025	8,694			39,719	Megga-Hardware/Self-performed work
08 11 16	Aluminum Doors and Frames				17,584	17,584	Boston Storefronts, Inc.
08 51 13	Aluminum Windows				21,238	21,238	Modern Aluminum Window, Inc.
08 71 00	Finish Hardware	37,000	19,668			56,668	Megga-Hardware/Self-performed work
08 81 00	Glass and Glazing				5,500	5,500	Able Glass Co.
09 21 00	Gypsum Drywall Systems				113,910	113,910	Eastern Drywall Systems Co.
09 30 13	Ceramic Tile				32,000	32,000	Western Ceramic Tile
09 51 00	Acoustical Ceilings				83,670	83,670	Capital Ceiling Systems
09 65 00	Resilient Flooring and Base				79,800	79,800	Top-Notch Floors
09 68 00	Carpet				112,000	112,000	Top-Notch Floors
09 91 00	Painting & Coatings				99,780	99,780	Commercial Painters, Inc.
10 11 00	Visual Display Boards				13,500	13,500	Office Interior Contractors
10 14 00	Identification Devices				3,500	3,500	Office Interior Contractors
10 21 13	Toilet Compartments				12,500	12,500	Office Interior Contractors
10 28 13	Toilet Accessories		3,542		3,500	7,042	Office Interior Contractors
10 44 00	Fire Protection Specialties		540		2,000	2,540	Office Interior Contractors
11 40 00	Food Service Equipment				123,490	123,490	The Kitchen Suppliers Co.

Table 5.2 (*Continued*)

Bid Date:	**July 20, 2023**	**New and Renovated Office Building**				**BASE Bid**	
Time:	**2:00PM**	**Boston, MA**				**Addenda;**	**#1, #2**
Sect.	**Description**	**Materials**	**Labor**	**Equipment**	**Sub**	**Totals**	**Remarks**
12 34 00	Manufactured Plastic Casework				29,750	29,750	New England Casework
12 21 00	Window Blinds				4,214	4,214	Clear-Vue Window Decor, Inc.
13 12 10	Exterior Fountains				34,000	34,000	Pete's Fountain Co., Inc.
14 24 00	Hydraulic Elevators				155,456	155,456	Uptown Elevators, Inc
14 42 00	Wheelchair Lifts				16,883	16,883	Uptown Elevators, Inc
21 00 00	Fire Suppression				65,000	65,000	Safety Fire Protection Co.
22 00 00	Plumbing				138,900	138,900	Best Plumbing Co.
23 00 00	HVAC				890,990	890,990	New England Pipe HVAC Co., Inc.
26 00 00	Electrical				292,300	292,300	Sparky's Electrical, Inc.
27 30 00	Voice Communications				19,455	19,455	Best Telephone and Data, Inc.
28 31 00	Fire Detection and Alarm				28,345	28,345	Sparky's Electrical, Inc.
31 11 00	Clearing and Grubbing				23,560	23,560	Super Sitework Contractors, Inc.
31 23 00	Excavation and Backfill				329,899	329,899	Super Sitework Contractors, Inc.
31 23 19	Dewatering				13,400	13,400	Super Sitework Contractors, Inc.
31 64 00	Caissons				234,500	234,500	Super Sitework Contractors, Inc.

(*Continued*)

Table 5.2 (*Continued*)

Bid Date:	**July 20, 2023**	**New and Renovated Office Building**				**BASE Bid**	
Time:	**2:00PM**	**Boston, MA**				**Addenda;**	**#1, #2**
Sect.	**Description**	**Materials**	**Labor**	**Equipment**	**Sub**	**Totals**	**Remarks**
32 11 00	Gravel Base Course				56,900	56,900	Super Sitework Contractors, Inc.
32 11 26	Bituminous Concrete Paving				93,450	93,450	Hot Stuff Paving Co., Inc.
32 13 13	Concrete Paving (Sidewalks)			1,277	5,785	7,062	Jones Concrete Finishing Co.
32 16 13	Bituminous Concrete Curb					–	Hot Stuff (included in 32 11 26)
32 17 23	Pavement Markings					–	Hot Stuff (included in 32 11 26)
32 31 00	Fencing and Gates				5,500	5,500	All State Fence Co, Inc.
32 80 00	Irrigation				29,555	29,555	Green Side Up Landscaping Co.
32 92 00	Turfs and Grasses				26,975	26,975	Green Side Up Landscaping Co.
32 93 00	Plants				13,245	13,245	Green Side Up Landscaping Co.
33 11 00	Water Distribution Piping				78,000	78,000	Super Sitework Contractors, Inc.
33 31 00	Sanitary Sewer System				156,000	156,000	Super Sitework Contractors, Inc.
33 41 00	Site Storm Drainage Systems				198,570	198,570	Super Sitework Contractors, Inc.
33 71 00	Electrical Service						Sparky's Electrical (included in 26 00 00)
	Sub Total	138,735	99,178	12,046	4,945,835	5,195,794	5,195,794
	All Risk Insurance					17,666	
	General Liability					41,566	
	Sub Total					5,255,026	
	Main Office Overhead	6.7%				352,087	
	Sub Total					5,607,113	
	Profit	8.0%				448,569	
	Sub Total					6,055,682	
	Performance and Payment Bonds	0.93%				52,146	
	BID					**6,107,828**	

Cost control can be broadly defined as the practice of analytically measuring and forecasting costs associated with a task or the entire project. It has the goal of minimizing those costs and any impact changes will have on cost during the life of the project. There are four basic aspects to project cost control:

1. Establish a baseline as a cost performance metric.
2. Measure actual costs against the baseline.
3. Accurately predict variations from the baseline cost.
4. Take corrective action(s) to minimize or eliminate the deviations.

The previously noted four steps are unsurprisingly similar to the schedule control discussed in Chapter 3 Pre-Construction Planning. In order to establish a baseline for comparing actual to estimated costs, the team must start with all of the available information about expected costs. This usually begins with the estimate. It is the first step in cost control since it provides a budgeted amount for the work. The cost control system is used to identify variations from this budget, which are then brought back in line with the budget. It then makes sense that the estimate and cost control system share information during the project life cycle.

Deconstructing the Estimate to Build the Baseline Budget

How the estimate is built is sometimes different than the way costs will be tracked. This is partly attributable to estimating practice and, to a lesser degree, how the costs are collected. Another factor in the mix is the plan to execute the work. Recalling Chapter 3 Pre-Construction Planning, occasionally work is estimated by one method and then during the planning phase, that method may be changed in favor of a new approach. Whatever the means and method selected to execute the work, the estimated value of the work in a stipulated sum contract does not change.

Tasks in a unit price estimate have an extended value (see Table 5.1) that must be analyzed, sorted, and separated into the appropriate "buckets," or *elements*, for cost control.

The process of breaking down the estimate is called *deconstructing* the estimate. The level of detail required during the tracking of costs is predicated on the level of control needed to manage the project. While the deconstruction process will vary from project to project, there are a few considerations that are consistently helpful in determining the level of detail needed:

- Tasks that are self-performed traditionally require a higher level of detail than work that is subcontracted. Work that is subcontracted has a built-in guard on cost control, which is the maximum amount of the contract. Unless a change order is issued, costs cannot exceed the contract price for the GC.

- Tasks that have long durations require a higher level of detail than short duration tasks. Productivity on tasks that extend over long periods can fluctuate as a result of boredom and complacency within the workforce.
- Tasks that are extremely labor-intensive (large workforce) require a higher level of detail as costs can escalate very quickly.
- A task that is complex in execution, or has numerous steps to completion, represents a higher risk for failure and, therefore, requires a higher level of detail for better control.
- Tasks that have never been done by the crew will require a higher level of detail to determine where, if any, inefficiencies occurred or variations from the estimate.
- Tasks in which we hope to collect or bolster historical data for future estimating may require more detail.

Prior to starting the deconstruction process, the project manager and project team must have a thorough understanding of the estimate. This includes assumptions made during the estimating process and any qualifications, alternatives, or substitutions that may have been accepted as part of the contract. Clearly the level of detail needed for work that is subcontracted with a fixed price is less than that needed if the work is self-performed. In fact, the GC is not privy to the actual costs incurred by the subcontractor, nor is it really necessary to know. Remember that the contract amount between the GC and the subcontractor *is* the cost of the work for the GC. This is what the GC will pay, irrespective of the actual cost to the subcontractor.

The Cost Breakdown Structure

The deconstruction process begins with the estimated cost that represents the contract amount. By definition, cost does not include profit. It is the only item in the estimate that does not have a corresponding payout by the contractor. This is the same starting place for developing a task list in the development of the CPM schedule. The as-bid estimate serves as the upset or maximum sum for the *original budget*. The original budget for a contract does not change. Additions or deletions from the original budget by approved change order are reflected in the updated budget, or *current budget*. Budgets should be updated at the same intervals that schedules are updated in order for cost and schedule to be synchronous. This provides the most accurate snapshot of project status and provides the latest information on which to base decisions from both a cost and time perspective.

The original budget is broken down into a *Cost Breakdown Structure* (CBS). The CBS is similar to the WBS introduced in Chapter 4 The Schedule, with the exception that the focus is on the cost model for the project. These sections can often follow the CSI MasterFormat breakdown by the division, noted earlier. The practicality of a logical CBS becomes evident each update period during the analysis of the project cost and progress. The CBS is composed of multiple layers or levels starting with a big picture or summary view and working its way down to the detailed costs. It is far easier to review costs at the summary level, and then focus on the details as necessary to analyze any variations. There are significant similarities between a WBS and a CBS, and ideally, they should be the same. As will be shown later in the text, the closer they are to being the same, the easier it will be to assign a dollar value to a task in the schedule.

It is not uncommon for a CBS to have multiple levels of breakdown, with the detailed *cost account* at the bottom. Here is an example of CBS for creating a cost baseline:

- Level 1: Division 06 00 00—Wood, Plastics and Composites
- Level 2: Section 06 10 00—Rough Carpentry
- Level 3: Sub-section 06 10 53—Miscellaneous Rough Carpentry

The Cost Account

The cost account is the lowest level of the WBS. Costs associated with a task are tracked in cost accounts. Cost accounts are also known by other names such as "detail accounts," "cost codes," and "chart of accounts." For the purposes of this discussion, all names of cost accounts will be used interchangeably. An example of a cost account from the CBS would be Level 3: Miscellaneous Rough Carpentry.

The Cost Element

A subset of the cost account is the *cost element*. The cost element is used for the separation of costs by type. Types of cost include material, direct labor, equipment, and subcontract. Frequently, labor and equipment usage can be tracked by hours expended in addition to dollars, and so these additional elements may be needed as well. If the work of miscellaneous rough carpentry cost account were self-performed, an example of the cost elements to be tracked would be:

1. Miscellaneous Rough Carpentry—Materials (dollars)
2. Miscellaneous Rough Carpentry—Direct labor (dollars)
3. Miscellaneous Rough Carpentry—Equipment (dollars)
4. Miscellaneous Rough Carpentry—Labor-hours (hours)
5. Miscellaneous Rough Carpentry—Equipment hours (hours)

The CBS, the cost accounts, and the cost elements all have a numbering system that allows for tracking, sorting, and reporting by computer. While the actual numbering system may vary by company or by software employed, the basic idea is the same. It is not uncommon and actually ideal to have the WBS and the CBS match numerically and employ CSI MasterFormat as the organizational structure. This practice will be discussed in more detail in Chapter 6 Integrating the Schedule and the Budget.

Table 5.3 is an example of the CBS, the cost account, and the cost element as it might appear in a project cost report.

The Project Cost Report provides insight into the cost elements. It compares actual costs with planned costs in a tabular format. The Budgeted Value column of the report is a derivative of the estimate. It can be traced to the task line(s) in the estimate. It indicates the categories; Materials, Direct Labor, Hours, etc. that the team wants to track. It also provides detailed costs as of a specific date called the *status*, *reporting*, or *data* date. The status date should be the same as the status date in the schedule.

Summary Cost Account

Upper-level managers frequently need summarized status updates where the actual details are less important than the big picture. This big picture reporting is prepared in a *summary cost account*. The summary cost account is defined as the compilation of cost accounts at a specific level of the CBS. For an upper-level manager who may have oversight responsibilities for 15–20 projects, it is far easier to analyze trends in a dozen summary accounts then to scan reams of paper representing five hundred separate cost accounts.

Cost control and cost accounting are very different in both concept and practice. Cost accounting is the tracking and recording of cost expenditures incurred on a project. It adheres to accepted bookkeeping/accounting practices without which a true and accurate status of the project may never be known. The structure of the accounting practice is intended to serve whatever requirements there may be with regards to capitalization, taxation, regulation, or contract requirements. Cost accounting will assume as

Table 5.3 Project Cost Report Line 06-10-53 Miscellaneous Rough Carpentry

Cost code/element	Budgeted value	This period	Job-to-date	Percent	Variance
06-10-53.10 Materials	$ 14,675.00	$ 7,655.14	$ 11,203.28	76.3	$ 3,471.72
06-10-53.20 Direct Labor	$ 85,655.00	$ 42,345.22	$ 65,955.22	77.0	$19,699.78
06-10-53.30 Equipment	$ 5,435.00	$ 4,641.45	$ 4,641.45	85.4	$ 793.55
06-10-53.40 Labor-Hours	1,540.00	760.90	1,185.80	77.0	354.20
06-10-53.50 Equipment Hours	65.50	54.50	54.50	83.2	11.00
06-10-53.60 Total Dollars	$ 105,765.00	$ 54,641.81	$ 81,799.95	77.3	$23,965.05

the basis the contract sum, including the profit component. Cost control, on the other hand, does not reflect the profit anticipated; it deals strictly in cost. While the distinction may be subtle in syntax, in practice, it has quite a different impact.

In establishing a cost control system, the goal is to identify and control those cost accounts and elements with the greatest potential to affect the final cost (positively or negatively) and with only summary level control on the remaining elements. Cost accounting makes no such distinction. It tracks all costs incurred.

One last note: The greatest variable in construction is direct labor cost. It is often referred to as the "wild card" because of all of the circumstances and conditions that can affect it. Thus, a contractor engaged in serious cost control will focus a laser-like intensity on labor: hours, productivity, morale, or any other contemporaneous factor since the labor costs can often be the determining factor in a successful project.

Other cost elements such as material and equipment can be reasonably predicted with a high degree of accuracy, especially if the project has been estimated correctly. It should also be recognized that equipment and labor have a direct correlation since most equipment needs an operator.

Project Cost Report

Project cost reports, also called *job cost reports*, are the lifeblood of the cost control cycle. Cost reports are produced by computer and as such have similar features. In general, a cost report tracks costs from two reference points; currently (specific period) and in total (cumulative). Cost reports describe in detail the actual costs incurred during a specific reporting period within individual cost accounts by cost element. This is called the *cost-this-period* subtotal. In addition to reporting by the period, the job cost report maintains a cumulative total of the costs incurred on the same cost element. This cumulative total is called the *cost-to-date* total. Comparing the current period to the cumulative total allows the project manager to spot trends that may be developing from period to period. Many types of reporting software forecast the cost at completion based on the current trend. While trends are subject to change from reporting period to reporting period, they provide a good indication as to the direction the cost of a task is headed. The trend used in conjunction with the progress report from the updated schedule provides the project manager with valuable information for decision making. Job cost reports should clearly show the actual cost performance as compared with the estimated.

Data entry on costs incurred is typically done as invoices are received. Payroll costs are posted on a weekly or biweekly basis with the reporting period to coincide with the update of the schedule. This is not true with all companies, as some project managers prefer to see a more frequent cost report than schedule report. Regardless of the reporting interval it is paramount that the status date on the schedule progress report and the job cost report are the same! Different dates on each can present a very distorted picture of what is happening with cost for the progress achieved.

The Budget as a Management Tool

The CPM schedule can be used to manage physical performance by providing information to the project manager to increase crew size, work longer hours, or add a second shift. In a similar manner, the budget acts as the maximum upset or cap to the financial side of the controls process. Once the baseline budget has been set, the project manager uses the budget as a cost metric for expenditures. As individual cost accounts show trends, the project manager can decide what action, if any, is appropriate. Negative trends may trigger an immediate investigation into the problem. In most circumstances, timely investigation coupled with a corrective measure can often reduce or reverse a negative trend. Trends can be positive, which indicates the work is progressing at less cost than anticipated in the budget. While positive trends do not typically require any corrective action, the project manager may investigate the reason for enhanced performance to see if it can be applied to improve the performance of another cost account.

Recasting the estimate into the budget can occasionally be a less than straightforward assignment. Multiple costs in the estimate are often combined into a single line item, which is best tracked individually for project control. Portions of combined work may also be performed by separate work crews on-site, which furthers the need for separating costs. For example, assume that the estimated cost for formwork (in SFCA) includes the cost to place the concrete as well. Since the formwork is erected and stripped by carpenters, and the concrete is placed by laborers, the project manager may choose to track these costs separately. In setting up the budget, without intimate knowledge of the cost of concrete placement, the project manager may have to assign a less than exact value for both the forming/stripping and the placement. This is a common and acceptable practice. The disadvantage is that costs for each category (forming/stripping and placement) may not reflect true costs for each scope of work. However, the combined costs are representative of the full scope of work.

An important side note is that we have discussed internal controls to this point. There are external control devices that should at least be introduced at this point. One in particular is the Schedule of Values (SOV). The SOV is the breakdown of costs for the purpose of requisitioning money from the owner/client. Since our industry is based on "a pay as you go" plan, there has to be an acceptable method of breaking up a large contract into smaller pieces for payment. Nobody waits till the end to get paid. This is the SOV. The most recognized SOV is the AIA G702 and G703 *Application for Payment* document published by the American Institute of Architects. It allows the contractor to again decompose the estimate into recognizable, measurable, tasks or groups of tasks that can be used to invoice the owner for work completed. It allows the contractor to draw down the contract sum in accordance with the progress of the work and its value. While this is slightly outside the strict sense of project control, it is related in a large way. As they track the cost incurred and progress achieved under project control guidelines, the PM team must ensure that they requisition a sufficient amount of money to pay the bills incurred, plus the commensurate amount of overhead and profit to maintain the accounting for the project in the black. This will be discussed in depth in Chapter 6 Integrating the Schedule and the Budget.

Summary and Key Points

Of equal importance to the integrated control process is the project budget. Since few projects, if any, have unlimited financial resources, tracking costs are an essential part of determining project status.

The estimate is deconstructed into a Cost Breakdown Structure (CBS) to arrive at the as-planned budget for entire categories of work as well as detailed tasks. The level of detail in the budget is directly related to the level of project control necessary to manage the project. CSI MasterFormat, used for the WBS in the scheduling process, is often employed to arrive at the CBS. The CBS should be representative of the WBS in the schedule. This provides continuity across both cost and schedule that provides for more accurate and timely information by which better decisions can be made.

The individual cost account for a task is composed of cost elements, material, labor, equipment, labor-hours, etc. The more detailed accounting of costs, the more information that is provided to the team. However, not all cost accounts require this level of detail especially if the work is subcontracted. In this case, the cost is the contract amount.

Much like the schedule progress update, the cost update can provide a clear picture of where the tasks or project stand. Costs must be accurate and timely and are generally included in a job cost report. This tool can be used to manage the cost side of the work.

The moral of the story is that the absence of a detailed unit price estimate may preclude accurate cost control at a detailed level. It may also not be a requirement of the team. Too much detail in a cost report can leave the project management team confused as to what was the goal of the tracking. It may render the data useless for any practical purpose of control.

In Chapter 6 Integrating the Schedule and the Budget, we will explore how the integration provides the necessary tools to manage the project.

Key points of this chapter are:

- Cost control represents the other half of the project control system.
- The breakdown structure for cost is the CBS which should match the WBS for the schedule.
- Costs follow the same structure as the technical data; CSI MasterFormat. This provides a level of detail that can represent a task in the schedule.
- To manage the budget, the PM and team need updated and accurate job cost reports that allows timely decisions based on accurate information.
- Reporting for cost and schedule parameters must be the same status date.

Chapter 5 The Budget: Questions for Review

1. Define the term "cost" and explain how it is affected by its frame of reference.
2. Specifications are qualitative in nature and set the acceptable standards that the quality of the project will be measured against. True or False?
3. Which type of estimating is the methodology by which bid documents are broken down into their incremental components for pricing?
4. The compilation of the estimated costs for a project becomes the budget for the project. True or False?
5. Explain what CSI MasterFormat is and how it is used in the construction industry.
6. What type of information is contained in Part 1 General section of a technical specification section?
7. By what methods are products contained in the technical specifications specified?
8. What is a cost element?
9. What is the purpose of the project job costs report?
10. Additions or deletions from the original budget by approved change order are reflected in the updated budget called?

CHAPTER 6

INTEGRATING THE SCHEDULE AND THE BUDGET

In Chapter 4 The Schedule, we reviewed the process of scheduling using the critical path method. We also discussed how a baseline schedule is used for time management in the project control process. In Chapter 5 The Budget, we reviewed how the estimate is used to construct a baseline budget for cost control. We have discussed the individual contributions of both to the project control process. Now let's integrate the schedule and the budget for the full impact of project control. It is the combination of both schedule and budget that allows the project manager the range of control needed to manage the work.

Let's assume that our team has created and accepted a schedule that will be used as the baseline for time management. That same team has deconstructed the estimate, created a Cost Breakdown Structure, and is now ready to integrate the two.

Schedule of Values

Before delving into assigning costs to tasks, there is a frame of reference issue that needs to be clarified. Previously, it was stated that the majority of contracts are performed under a stipulated sum delivery method. This delivery system is based on providing work of a definitive scope for a fixed price. Under this methodology, the stipulated sum is based on a proposal or bid that eventually matures into a contract with a fixed price. This process may include negotiation to arrive at the contract sum. This delivery methodology is the framework for the following examples. Within that contract price is the profit that the contractor has included in the estimate. This estimate upon award of the contract, will become the basis for the budget we will use in cost control.

We also defined that cost is relative to the frame of reference it is viewed from. For the owner, who will pay the contractor, cost includes the contractor's profit, also known by its more politically correct term—"fee."

A major part of the project control system is translating the progress of the work into earnings or *receivables*. It is receivables or cash flow that

Project Control: Integrating Cost and Schedule in Construction, Second Edition. Wayne J. Del Pico.

allow the contractor to pay direct costs incurred during the performance of the work, including the overhead to manage the process. Since most contractors have neither the risk tolerance nor the financial stamina to wait for one payment at project completion, the work is billed as it progresses. The process by which the contractor bills or requests payment is called *requisitioning*. Requisitioning is the periodic billing for the value of work that has completed in that period. The most common billing period is the month or typical 30-day business cycle.

Chapter 5 The Budget discussed the assigning of a Cost Breakdown Structure or CBS and the tracking of costs from the contractor's perspective. The CBS and tracking of costs are highly proprietary to the contractor's method of operation and business model and are very rarely, if ever, shared beyond the contractor's most trusted employees.

However, because the owner is essential in the payment process and work is requisitioned periodically, it is necessary for the contractor to establish values for tasks or groups of tasks to serve as the basis of the billing. Again, it is a separation of the work items that comprise the contract by their respective costs.

The industry term for the document that breaks the contract down into billable values for work is called the *Schedule of Values* (SOV). This was introduced briefly in Chapter 5 The Budget. The SOV is the decomposition of the contract into tasks or groups of tasks that have a dollar value assigned to them. The total of the SOV must equal the contract price including the fee or profit.

The SOV is used in the measurement process for determining what work has been accomplished or earned. This will be discussed in detail later in the text. For now, the focus is just on the Schedule of Values. It should be noted that the word "schedule" in the name is misleading. The SOV is on the financial side of control and not on the time management side. The definition of the word schedule, in this application is "a written statement of details in a tabular format." It has nothing to do with time.

There are numerous schools of thought surrounding the breaking down of the work for the SOV. The most common of which is the "more detail–less detail" argument. Some contractors subscribe to the belief that the less detail, or less breakdown, of the contract the better. The theory is that it offers a smaller window to the owner into the inner cost breakdown of the work. The disadvantage with less detail is that it provides a less factual basis for determining the value of the work that has been performed. The value of the work is more subjective to experience or opinion, which can be a source of dispute. Since most owners would prefer to err on the side of caution and not overpay the contractor, these disputes can often lead to a reduction in cash flow for the contractor and subcontractors.

The opposing argument is that more detail is better. While this does allow a more exposed view to the breakdown of the contract value, it also provides a more factual basis for determining the value of work completed with less interpretation or opinion by the individual reviewing and approving the requisition.

As an example, consider the following two breakdowns for the electrical work on a project:

Less Detail	
Electrical	$200,000.
and	
More Detail	
Permits	$ 5,000.
Underground Service	$ 15,000.
Panels	$ 20,000.
Branch Wiring	$ 20,000.
Raceways	$ 10,000.
Power Wiring-Equipment	$ 40,000.
Interior Lighting	$ 75,000.
Exterior Lighting	$ 10,000.
Closeout	$ 5,000.

While both examples total the same $200,000 value for the electrical work, the more detailed breakdown provides more information for determining the value of work completed. The less detailed method provides little information, which will force the reviewer to substitute opinion or experience in determining value of the work completed.

Another common debate is whether fee should be a separate line item or distributed across all items as a percentage of each task. Either one is acceptable and is typically dictated by the contract or owner's preference. This is the same with the distribution of home office overhead which can be included as part of the fee. Many owners believe that if the fee is a separate line item, there is a more accurate breakdown of the costs. With the distribution of fee across all items, there is a belief by owners that some items, especially those completed early on in the project, will carry a heavier percentage of profit. The concern is that the owner will pay more than the value of the work in place, and the contractor will be overpaid. This practice of placing a greater percentage of profit on the earlier items is called *front loading*. Most contractors practice some degree of front loading with the justification that it is intended to offset some of the financial burden caused by the retainage held by the owner. There is a certain amount of truth in this statement.

In tough economic markets with contractors accepting projects with profit margins in the single digits and retainage at 5 percent or 10 percent, it means that the contractor may have to wait till the project is complete and paid for 100 percent to realize a profit or even full overhead. This puts a disproportional amount of risk in the contractor's court and would be tantamount to the contractor providing additional financing for the owner. While it is impossible to front load the SOV sufficiently to offset all of the retainage, excessive front loading should be discouraged whether it is recognized by the owner or not. Contractors that practice excessive front loading often are left with insufficient funds to pay direct costs at the end of the project due to the devaluing of tasks that occur near the end. It is also considered a poor management practice by the sureties that provide the performance and payment bonds on projects. Sureties are at their greatest risk for failure by their principals near the end of the project. To exacerbate this risk by reducing needed cash flow is both dangerous and unnecessary. For the purpose of examples in this chapter, the author has chosen the method of distributing overhead and profit across all tasks.

As will be seen in the next section, there is a direct relationship between the project baseline schedule and the Schedule of Values.

Matching Cost Values to Tasks

The first step in the integration process is to ensure that there is a direct relationship between a task in the schedule and the line item that represents it in the budget. Simply stated, the task in the schedule must have a corresponding dollar (or labor-hour) value assigned to it. This value must be correct and accurately represent the cost of the work to be performed. The matching of the task to its realistic cost is essential to accurate project control. Without the correct assignment of dollar value to a task during the creation of the Schedule of Values, the project manager will never have a clear and accurate snapshot of project status for cost or time.

Matching costs to tasks in the SOV is nothing more than assigning the budgeted dollar value including the task's respective share of overhead and profit to the task. For example, consider the task of setting of precast concrete curbs on a project. Assume the task to be self-performed versus subcontracted. A look at the baseline budget for this task might appear like Table 6.1. The numbers in Table 6.1 originate from the detail sheets of the estimate, both quantities and prices for each of the three tasks that comprise the precast concrete curb task.

The cost account for this task would read as 32.16.13.26: Precast Concrete Curbs with a total cost of $62,340.77. Note the cost account code employs CSI MasterFormat as the numbering system. In addition to the CSI MasterFormat, a contractor would most likely add a project number for more definition and ease of reporting.

Table 6.1 Budget for Precast Concrete Curb

				Material		Labor		Equipment		Total
Section	**Description**	**Quantity**	**Unit**	**Unit cost**	**Total**	**Unit cost**	**Total**	**Unit cost**	**Total**	**Total**
32.16.13.26	Precast concrete curbs									
1.01	Fine grade and compact excavation for curbs	2450	SF	$ 1.45	$ 3,552.50	$ 1.21	$ 2,964.50	$ 0.58	$1,410.30	$ 7,927.30
1.02	Furnish and install precast concrete curbs	2450	LF	$13.39	$32,805.50	$ 5.85	$14,332.50	$ 1.09	$2,670.50	$49,808.50
1.03	Back up curb with 2000 psi concrete	26	CY	$98.12	$ 2,551.45	$12.88	$334.92	$66.10	$1,718.60	$ 4,604.97
	Precast concrete curbs total				$38,909.45		$17,631.92		$5,799.40	$62,340.77

The cost elements set up for tracking purposes would be as follows:

32.16.13.26.1: Precast Concrete Curbs–Materials–$38,909.45

32.16.13.26.2: Precast Concrete Curbs–Labor–$17,631.92

32.16.13.26.3: Precast Concrete Curbs–Equipment–$5,799.40

32.16.13.26.4: Precast Concrete Curbs–Totals–$62,340.77

While this budget may be suitable for tracking costs on the task, it would be less than adequate for the requisitioning of work completed. The previous number is "raw" or unburdened and does not contain markups for home office overhead and profit. These are typically added in the recapitulation phase on the Estimate Summary sheet. Assuming a combined markup for home office overhead and profit (O&P) of 10 percent, the same line item in the SOV would be:

$$\$62,340.77 + 10\% \text{ for O \& P} = \$68,574.85 \text{ or } \$68,575$$

The dollar value with the markup for overhead and profit is called the *scheduled value* for the task. This means that when the task is 100 percent complete the contractor will have earned $68,575.00. This number represents not only the cost to the contractor, but the incremental home office overhead and profit that have been assigned to the precast concrete curb task in the estimate. (This example has intentionally omitted any front loading for a clearer illustration.)

Now consider the schedule portion of the integration. Assuming that our contractor has self-performed this work in the past, there should be historical data available from prior projects which provides a baseline for productivity. Based on historical cost data, the production rate for our crew to install the curb is approximately 410 LF per day complete. This has been tracked over numerous other curb installation projects. Calculating the duration (D) of this task using the daily production rate method and the total quantity (Q) of 2,450 LF from the estimate (D), would be:

$$\text{D} = \text{Q} \div \text{DO}; 2,450 \text{ LF} \div 410 \text{ LF / day} = 5.97 \text{ days or } 6 \text{ days}$$

Based on this analysis, it is easily concluded that the task of furnishing and installing precast concrete curb for our sample project has a duration of six days[1] in the schedule and the budgeted value of the work (without overhead and profit) when complete will be $62,340.77. These two parameters are called *baseline values*. They represent the planned values for both time and dollars that will be expended when complete and will be tracked during performance. The requisitioning of this task will be compared to these performance metrics but will include the overhead and profit bringing the total to $68,575.

[1] It is a generally accepted practice to round schedule days up to the nearest full day.

In this scenario the cost that will be tracked for job cost purposes is represented by the estimated values in Table 6.1 totaling $62,340.77. In addition, another $6,234.23 will be added to the total to represent the increment amount of overhead and profit that will be billed to the owner for the task when it's complete. The total with overhead and profit of $68,575. will represent this work in the Schedule of Values.

It is further known from our cost account, 32.16.13.26: Precast Concrete Curbs, that the direct cost of this task based on our cost elements is a total of $62,340.77. From this information the project manager can conclude:

- If the direct costs to perform the work (materials, labor, and equipment) are $62,340.77, then the cost proceeded exactly as planned assuming each element is as estimated.
- If the costs total *more* than the $62,340.77, the work has cost more to perform than expected and money has been lost on the task.
- If the costs total *less* than $62,340.77, the work has cost less than expected to perform and profit from the task will be larger than expected.

We can conclude from the planned duration that:

- If the work is 100 percent complete in six days, the work was performed as planned and the schedule proceeded exactly as planned.
- If the work takes longer than six days, the work was performed slower than expected and scheduled time was exceeded.
- If the work takes less than six days, the work was performed faster than expected and we have decreased the scheduled time.

It should be noted that both cost and time are predicated on the means and methods that were the basis of the plan as well as the crew to perform the work. Once either parameter changes, crew size or the planned means and methods, our performance data is not representative of our plan. Also, the tracking of costs may include the incremental overhead and profit. This is the decision of the accounting professional and company policy. Our example has intentionally omitted the overhead and profit for clarity.

Planned vs. Actual Values

It is important to consider what other information can be derived from the integration of cost and schedule. In order to use this information to both manage the work and forecast performance, one must have an understanding of how work progresses and how value is earned. There are some assumptions that must be adhered to:

1. All tasks are being performed by experienced professional tradespersons.

2. The tradespersons have the appropriate materials, tools, equipment, and training to perform the task.
3. The crew assigned to perform the task is the optimal crew. All crew members are contributing to the production.
4. Work is being performed continuously without any interruptions or delays.
5. The workday remains a constant eight hours per day.

Consider how any task is performed. To start the task the crew assembles, receives direction, or reviews plans for scope, sets up materials and equipment with possibly some layout, and then proceeds to perform the actual task. Frequently the work starts off slow, until all crewmembers (or the individual) performing the work become comfortable with the process. This is referred to as the *learning curve* for the task. Learning curves vary considerably with the complexity of the task being performed. As the crew repeats the task over multiple cycles, their productivity improves until such time as they have reached the maximum productivity achievable for the crew performing that task. Near the completion of the task, productivity may wane, as some of the incomplete items are finished, or setups to perform the task are broken down, until finally the task is complete. This is called the *wrap-up* phase of the task. Figure 6.1 illustrates a task life cycle.

With the exception of mastering the learning curve at the beginning of the task and the wrap up at the end of the task, the work is performed fairly uniformly. That is to say that the production rate remains fairly constant over the majority of the duration. Because the construction industry recognizes the distinct phases in the performance of a task, it averages productivity across the full duration of the task. The logic is that the average more closely represents the productivity of the task than the productivity from any one specific phase. A more thorough discussion of productivity will be conducted in Chapter 10. If productivity is the average work accomplished per unit of time, and time is measured in workdays of eight hours, then the average daily output can be calculated with relative ease. The average daily output can be defined as the average daily amount of work produced over the duration of the task. (Both the Daily Production Method and the Labor-hour Productivity Method discussed previously are based on average productivity over the life cycle of the task.)

Expanding this concept further, the project manager should be able to forecast productivity based on the average daily output. Applying this concept to the precast concrete curbing example detailed earlier, the project manager can forecast how much curb has been installed at any day along the six-day duration of the task.

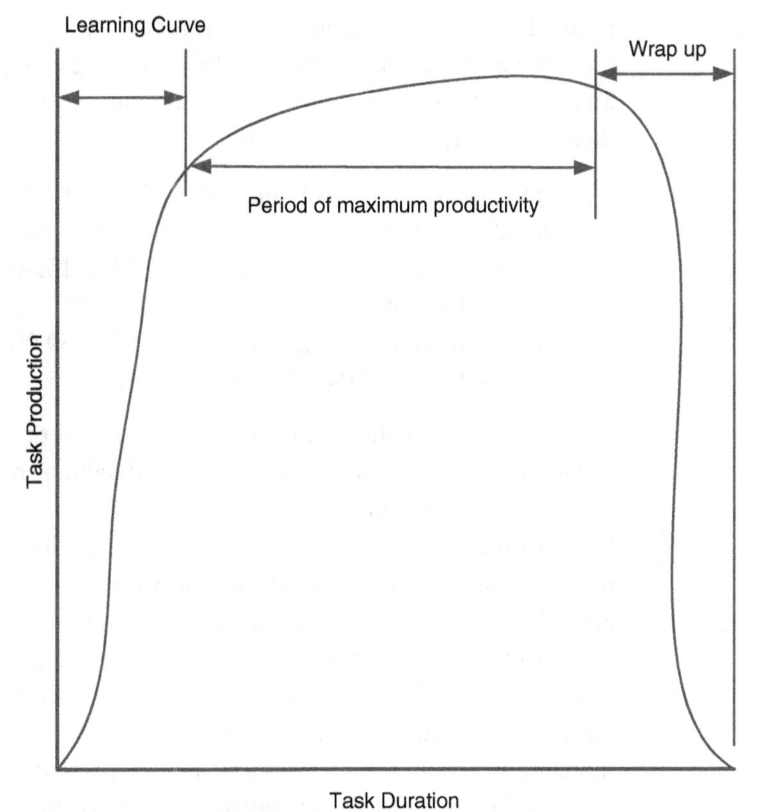

Figure 6.1 Task Life Cycle

If the average daily output of 410 LF per day is used, and we have a total of 2,450 LF, then:

- Day 1—410 LF is installed for a total of 410 LF
- Day 2—410 LF is installed for a total of 820 LF
- Day 3—410 LF is installed for a total of 1230 LF
- Day 4—410 LF is installed for a total of 1640 LF
- Day 5—410 LF is installed for a total of 2050 LF
- Day 6—400 LF is installed for a total of 2450 LF[2]

The project manager could use this information to determine what he or she could expect to see when visiting the site or when reporting on anticipated progress. For example, if the project manager were to visit the site at the end of the day on Day 4, he or she should expect to see approximately 1,640 LF of precast curb installed if the work is progressing as scheduled.

Combining both the scheduled value of the work in dollars from the Schedule of Values with the anticipated progress predicted in the

[2] The quantity has been adjusted for the remaining available quantity on Day 6.

schedule, the project manager can forecast earnings from this task at any point along the schedule. In turn, the project manager could anticipate the amount of money the contractor would earn as the work progressed by distributing the value at each day:

- Day 1—total of 410 LF and earned $11,471.80
- Day 2—total of 820 LF and earned $22,943.60
- Day 3—total of 1230 LF and earned $34,415.40
- Day 4—total of 1640 LF and earned $45,887.20
- Day 5—total of 2050 LF and earned $57,359.00
- Day 6—total of 2450 LF and earned $68,575.00[3]

The scheduled value of the task distributed over the duration of the task is called the task's *planned value*. Planned value is determined by assuming a pro rata distribution of dollar value over time and that the work will be performed reasonably uniformly for the duration of the task. In reality, the performance of the work may be substantially different from what actually occurs during execution. This integration of scheduled value with anticipated duration is the foundation of practical project control. This not only allows the project manager to plan the work to be accomplished, but can also predict *when* the money will be earned. The dollar value of the work scheduled to be accomplished in a specific time period is called *Budgeted Cost of Work Scheduled* (BCWS) or in new parlance *Planned Value* (PV).

Any project manager with a modicum of experience knows that work doesn't always go as planned. In fact, it rarely, if ever, proceeds *exactly* as planned. However, the relationship between schedule and value remains constant as the work is executed.

Revisiting the example of the installation of the precast concrete curb, at the end of Day 4 our project manager visits the site and determines that the progress is behind. The total curb installed to date at the end of Day 4 is 1500 LF instead of the 1640 LF that was anticipated by the schedule. As a result, the dollar value earned is commensurate with the progress and can be calculated:

$$1{,}500 / 1640 = .915; \text{ or } .915 \times \$45{,}887.20 = \$41{,}986.79$$

This new value represents the dollar amount earned based on the actual performance. It is called the *earned value* of the task and is a key component in assessing progress. When it is measured in terms of the planned value, it is called the *Budgeted Cost of Work Performed* (BCWP) or in new terms *Earned Value* (EV). The BCWP is the actual

[3] Adjusted for reduced quantity.

dollar amount earned based on what was scheduled to be earned. It is this relationship between the BCWP and the BCWS over multiple reporting periods that defines a *trend* or repetitive pattern in performance. The BCWS and the BCWP are two-thirds of the trinity called Earned Value Management.

The last step in this process is to analyze what it *actually cost* to achieve the earned value. Based on the estimate converted to the budget we have the anticipated cost of $62,340.77. This was explained in a previous section of this chapter. This means that when the work is 100 percent complete, the costs realized should be a maximum of $62,340.77. This is called the *Actual Cost of the Work Performed* (ACWP) or *Actual Cost* (AC). The cost of the work, especially for a task that lasts only six days would most likely be analyzed and compared when the work is complete. The reason is most likely that the accumulation of the costs that would be posted to the job cost report would not be uniform. Accounting might receive an invoice for the precast curb material before the work started, and the invoice for ready mix concrete well after the work was done, payroll for the labor portion might lag by a week or more. An analysis of the actual cost at any time in the six days might reveal an erroneous amount that could not be tied directly to earned value.

Actual cost and its relationship to BCWS and BCWP will be explored further in Chapter 7 Calculating and Analyzing Progress.

Summary and Key Points

Combining costs with schedule allows the project manager a new window into the status of the project. The direct relationship between progress and the amount of money earned remains constant. The project manager can use the integrated information to predict future performance in terms of schedule as well as for cash flow management. It can also be used to predict when the work will be completed if production is less than planned.

As the basis for payment the project manager creates the Schedule of Values (SOV) that reflects the full value of the work including overhead and profit. It is the SOV that will be used along with the schedule to establish the value of the work the contractor will be paid. It is critical that the individual items in the SOV represent real, not inflated values to obtain a genuine status of the cost model for the project.

In the execution of a task, the production is assumed to be linear and as a result the planned value of the work based on the time in the schedule can be predicted. This is called the BCWS. It distributes the work equally over the life of the task. Since the planned value doesn't always go as anticipated, the second step is to determine what was actually accomplished. This is called the BCWP. It measured the work complete and the value of it. The last step is to determine what it cost to earn the BCWP.

Chapter 7 Calculating and Analyzing Progress will explore the measurement process and the last component to the Earned Value Management system further.

Key points of this chapter are:

- It is essential to integrate both the schedule and cost of a task to understand the full status of the task.
- The Schedule of Values is a tool used by the contractor to break down the cost of the work so that it can be requisitioned from the owner/client.
- The Schedule of Values represents the full value of the contract when executed and the baseline budget represents the maximum anticipated cost of the work when complete.
- The difference between the baseline budget and the Schedule of Values for the same task is the overhead and profit.
- The schedule and the budget can be used to determine the progress of a task and what money has been earned by that progress. It can also be used to forecast completion and cost.
- The last step in the analysis is to determine what it actually cost to earn the value of the completed work.

Chapter 6 Integrating the Schedule and the Budget: Questions for Review

1. What is a Schedule of Values and what does it represent?
2. The full value of a line item in the Schedule of Values represents the value of the scope of work at 100 percent complete. True or False?
3. The practice of placing a greater percentage of profit, or inflating the value of tasks, occurring early in the project is called?
4. What are the five values that must remain constant in order to compare actual to planned value?
5. The Schedule of Values is used in determining the value of the work that has been completed in a specific period. True or False?
6. The scheduled value of the task distributed over the duration of the task is called?

7. The dollar value of the work scheduled to be accomplished in a specific time period is called?

8. The BCWP is the actual dollar amount earned based on what was scheduled to be earned. True or False?

9. Combining the scheduled value of the work in dollars from the Schedule of Values with the anticipated progress predicted in the schedule, the project manager can forecast earnings from this task at any point along the schedule. True or False?

10. Average daily production can be used to forecast future production if conditions remain unchanged. True or False?

CHAPTER 7

CALCULATING AND ANALYZING PROGRESS

A crucial part of the project control process is the accurate measurement of the progress of the work, without which it would be impossible to calculate the status of the project accurately. It is the measurement of progress for each individual item, however dissimilar from the next, that summarizes the overall progress of the project.

Analyzing progress and its monetary value is part of the tracking process and occurs during the execution of the work. The tracked progress is then compared to the anticipated progress, both cost and time, to determine if the project is achieving its intended goals on each front. However, it has little chance of success if the appropriate baseline controls for costs and schedule have not been established or if the measurement of the actual progress is inaccurate. Building on Chapter 6 Integrating the Schedule and the Budget, this chapter will explore ways in which to measure the units of completed work and apply it to the calculation of monetary earned value.

Why an Accurate Schedule of Values?

We have used the phrase "measuring progress" numerous times to this point, but what are we actually referring to? We are referring to measuring the work that has been completed since the last Application for Payment. The work is the individual tasks that have been assigned a value (scheduled value) when complete in the Schedule of Values that make up the Application. As the contractor completes the individual tasks that comprise the work, and the project moves toward Substantial Completion, the contractor is requisitioning, or "drawing down" the contract funds to pay for the work that has been done during the period leading up to the status date. Remember that invoicing in construction is broken into progress payments based on what has been completed in the pay period. The goal is that the earned valued must be sufficient to pay the actual costs incurred to perform the work. Without sufficient money requisitioned from the owner, the contractor will be required to augment the shortfall of funds from the

Project Control: Integrating Cost and Schedule in Construction, Second Edition. Wayne J. Del Pico.

contractor's own cash resources to pay the bills incurred. Most contractors would consider this an unacceptable practice.

Therefore, it is very important that the scheduled values for the work, which is the basis of the Application for Payment, is prepared with care. In a perfect world, the line item dollar amount in the Schedule of Values is the actual cost plus the commensurate overhead and profit for the task. This assumes that the overhead and profit are distributed over each task and not a separate scheduled value. When the line item is requisitioned at any point along its duration, the contractor will receive the cost of the work *plus* the incremental portion of the overhead and profit for the task. This is the most accurate way to determine if the task is ahead, behind, or as planned financially.

When the tasks are front loaded, especially to excess, the contractor receives more than the task is worth in comparison to the costs incurred to perform it. While most contractors would find having a little extra of the owner's money attractive, the excess funds received will signal an incorrect (too large) amount of profit for the task. Often this is recognized by the accounting professional and an adjustment is made. However, it is not always caught and will contribute to an incorrect Profit/Loss statement for the period. At the other end of the spectrum, if the value is too low in the scheduled values due to incorrect or hastily undervaluing the task, then the money drawn will be insufficient to pay the bills incurred for the task.

Measuring Performance

The application of sound measurement techniques during the analysis of earned value makes it possible to determine the status of the progress accurately. It also has the distinct advantage of providing an early warning sign for projects that are in trouble. As a first step in the Earned Value Analysis, the project manager must accurately calculate the units of work that have been accomplished. This is done at regular intervals in the project, typically monthly. There are generally six recognized methods for calculating progress on a construction task. These methods range from straightforward mathematical calculations to subjective ones based on experience.

Units Completed

The first method is straightforward and applies to types of tasks that involve repeated production of easily measured units of work. These tasks require approximately the same amount of time or effort to complete and rarely include subtasks. If subtasks are involved, they are typically performed simultaneously with the prime task. The placing and finishing of a concrete slab illustrate the type of work with subtasks that are handled simultaneously such as placing a vapor barrier, setting welded wire fabric, etc. Despite the subtasks, progress is typically calculated on the area of finished slab or by the cubic yard placed.

Another good example of a task whose progress is easily measured in terms of units completed is the installation of 2' × 4' LED light fixtures dropped into an acoustical ceiling grid. If a building had 200 light fixtures to install and 100 of them were installed, the percent complete would be 100 ÷ 200 = .50 or 50 percent complete. The percentage complete is based on a physical counting of the number of light fixtures installed. This can be applied to a large majority of construction tasks and requires no experience-based judgment that can be challenged.

Incremental Milestone

The second method is best used for a cost account that is comprised of subtasks that must be completed in sequence. It is called the *incremental milestone* or *steps method.* As an example, consider the forming and placing of a concrete foundation wall. There are a series of steps or operations that are performed in sequence, and each one contributes to the final product. The operations to complete the task include the following:

- Lay out foundation walls on footings.
- Erect inside wall panels and brace.
- Set horizontal and vertical rebar in formwork.
- Erect outside wall panels and brace.
- Place and finish concrete in formwork.
- Strip and remove formwork.
- Break ties and patch voids/holes.

For each of these operations, there is a budgeted amount of time (labor-hours) that will be expended by the crew to accomplish the work. As each operation is completed, a mini-milestone is accomplished and represents a percentage of the total installation of the foundation wall.

For example, here are percentages of the cumulative total applied to each operation based on labor-hours:

- Lay out foundation walls on footings – 32 hrs = 10.0%
- Erect inside wall panels and brace – 88 hrs = 37.5%
- Set horizontal and vertical rebar in formwork – 40 hrs = 50.0%
- Erect outside wall panels and brace –100 hrs = 81.0%
- Place and finish concrete in formwork – 20 hrs = 87.5%
- Strip and remove formwork – 30 hrs = 97.0%
- Break ties and patch voids/holes –10 hrs = 100.0%
  Total –320 hrs

The percentage chosen to represent each operation is based on the labor-hours expended as a ratio to the total. Each step is cumulative and is equivalent to having completed that percentage listed of the total installation. In order to use the incremental milestone process, there would have to be an agreement in advance as to the appropriate percentages of each of the subtasks. Also, calculating the breakdown of the intermediate tasks and the money allotted to each can be time-consuming.

Start/Finish

The Start/Finish method has its greatest appeal for those tasks that lack readily definable intermediate milestones or for which the time required for each operation is difficult to estimate. It focuses on only the start and finish of the task. This method has its greatest success with tasks that are fairly short in duration. These tasks have an easily discernible start and finish date, but status updates in between are not always possible. Classic examples include testing services such as load tests on electric panels, flushing and cleaning of piping, and similar tasks.

The Start/Finish approach assigns a percentage of the total value at the start of the task and the remaining percentage when the task is complete. For example, what is called the "50/50 rule" works on shorter duration, lower-value tasks. Under the 50/50 rule, 50 percent of the value of the task is earned upon commencement. At completion of the task, the remaining 50 percent is earned. Other "rules" include the 20/80 rule for slightly longer duration and higher-value tasks. For very short tasks, the 0/100 rule is often applied. That is, there is no earned value until the task is 100 percent complete. This is often applied to testing tasks that have little if any value until the results of the test are available. Consider a HVAC test and balance report. Regardless of the agreed-upon percentage, value is earned at only the start and the finish of the task. The applicable rules to use and the respective percentages are often the results of a negotiation between the contractor and the owner or design professional.

Cost Ratio

The Cost Ratio approach is applicable to tasks such as project management, general conditions, quality controls, contract administration, and even profit and overhead (if a separate scheduled value). This approach works best on tasks that involve long durations and are continuous over the life of the project. They are budgeted on mass allocation of dollars versus labor-hours of production. The Cost Ratio approach allows the contractor to earn value commensurate with the overall percent of project completion. For example, under the Cost Ratio approach, if the overall progress of the project was determined to be 42 percent, then the contractor would have earned 42 percent of the overhead and the fee. This is often applied to project overhead costs despite the fact that the expenditure of project overhead is not uniform across the duration of the project. The Cost Ratio approach can often be refined to make it more reflective of a uniform project overhead. For example, items or tasks such as prepaid insurances, a P&P bond, and project closeout can be listed in the Schedule of Values as individual items that can be requisitioned for using a Start/Finish method. This can leave the routine costs subject to the Cost Ratio approach.

Experience/Opinion

This is the most subjective approach with often no factual substantiation to support the results. It is used for minor tasks only where a more definitive basis cannot be used. It is most often used for operations such as dewatering, frost removal/protection, or constructing support facilities. It is typically relegated to tasks with small dollar values. It should come as no surprise that this method can be the source of substantial disagreement between the contractor and the owner or architect. Contractors tend to be "robustly optimistic" in their assessment as to what percentage has been completed under this approach. Design professionals and owners recognize the practice and often undervalue the work completed and earned value as a means of balancing the percent complete.

The Experience/Opinion approach should be used sparingly and sidelined for a more definitive approach.

Weighted or Equivalent Units

This method has its greatest appeal where the task under control has a long duration and is composed of multiple subtasks with dissimilar units of measurement. The erection of a structural steel package is a good example since it has several subtasks to achieve the finished product, each with a different unit of measure, all requiring resources and labor-hours to complete. In this case each subtask is weighted according to the estimated level of effort (in labor-hours) or by dollar value that is dedicated to each subtask. This weighted value is then transferred into the predominant unit of measure for the task: in the case of structural steel, tons. As a quantity of each subtask is completed, it is converted to equivalent tons to arrive at the percentage complete of the prime task, the erection of steel. Table 7.1 is an example of the Weighted or Equivalent Units method.

A variation of this approach utilizes equivalent units for each operation. In the previous example, each operation would be assigned a unit of measure that is equivalent in tons, even if that was not the same unit of measure in the estimate.

Table 7.1 Example of Weighted Method

Weight (tons)	Subtask	Unit of measure	Quantity total	Equivalent steel wgt.	Quantity to date	Earned tons*
0.02	Check foundation bolts	EA	200	10.4	100	5.2
0.02	Shim	%	100	10.4	100	10.4
0.05	Shakeout	%	100	26.0	100	26.0
0.06	Columns	EA	84	31.2	74	27.5
0.10	Beams	EA	859	52.0	250	15.2
0.11	Cross-braces	EA	837	57.2	0	0.0
0.20	Girts & sag rods	bay	38	104.0	0	0.0
0.09	Plumb and align	%	100	46.8	5	2.3
0.30	Connections	EA	2977	156.0	74	3.9
0.05	Punchlist	%	100	26.0	0	0.0
1.00	**Steel**	**TON**		**520.0**		**90.5**

* Earned tons to date = (Quantity to date) (Relative weight) (520 tons)/(Total Quantity).
Percent complete = 90.5 tons,/520 tons = 17.4%.

It is not uncommon to have a mix of these six methods on a single Application for Payment when trying to determine the value of the entire project. Often the simplest, more straightforward approach to the calculation of progress works best.

Earned Value

Having reviewed the accepted methods for measuring progress on individual tasks, the next challenge is to determine the percentage of completion for the project as a whole. The process for determining either the progress on an individual task or the project as a whole is called *Earned Value Analysis*. Earned value is linked to the budget values used in the Schedule of Values. It is typically expressed in dollars (or labor-hours) since they are the only common denominator among the variety of cost accounts found on a project.

The Earned Value Analysis starts with the time-phased costs that were addressed in the discussion of the SOV, which provides the project baseline that is the anticipated or planned value of the work scheduled in accordance with the budget.

Earned Value Analysis can be applied when budgets are both fixed (stipulated sum contracts) and variable (cost plus contracts).

Fixed Budget Approach

The Fixed Budget approach is used with stipulated sum contracts in which the scope of work is tied to a fixed dollar amount. The fixed budget amounts are the basis of the cost accounts. As in the discussion of the Schedule of Values (see Chapter 6 Integrating the Schedule and the Budget), the Fixed Budget approach utilizes each individual line in the SOV as the maximum amount that can be earned for that item when the work is 100 percent complete. It establishes a direct relationship between the percent complete and the SOV maximum. The relationship for a single task is expressed by the following formula:

$$\text{Earned Value (EV)} = \text{Percent Complete} \times \text{Maximum Budget}$$

As demonstrated by the formula, the higher the percentage of completion, the higher the earned value until the maximum budget amount has been earned when the task is 100 percent complete. The contractor can never earn more than the maximum budget amount for the task since mathematically the work can never be more than 100 percent complete. Consider the task of installing the light fixtures from earlier in this chapter. If the progress was measured at 50 percent complete and the total value for all 200 light fixtures was $35,000 when complete, we could determine the earned value at 50 percent as follows:

$$\text{EV} = (0.50) \times \$35{,}000 = \$17{,}500$$

Since progress in all cost accounts can be calculated using the same formula, the products can then be summarized to arrive at the total progress of the project as a whole:

$$\%\text{Complete} = \text{Sum of EV of all tasks} \div \text{Maximum Budget value of all tasks}$$

The Fixed Budget approach derives its accuracy from the correct distribution of costs within cost accounts (see "Why an Accurate Schedule of Values?" section above). Should the initial distribution prove incorrect due to poor estimating practice, incorrect quantities, or even unrealistic productivity assumptions, the project manager should redistribute the available budget amounts to match the recently acknowledged work conditions or the corrected requirements. If the project has been correctly estimated, the redistribution will provide a more accurate view of progress. If the cost account is underbudgeted, it will become evident fairly quickly by showing a higher percentage of completion than visual inspection with measurement will support. Cost overruns that arise on properly distributed cost accounts have the advantage of helping provide more accurate historical data for estimating the same task in the future.

Variable Budget Approach

The Variable Budget approach is particularly suited to projects with a nonfixed or cost-plus delivery methodology. On projects with Cost-Plus Fee contracts, the maximum upset is estimated based on the best information available at the time, but is subject to change as the full scope becomes known or the contract documents are completed. In contrast to the Fixed Budget approach, which is constrained by the fixed budget for the entire project, the Variable Budget approach makes use of the changing budget.

In much the same manner as the Fixed Budget approach, each task is assigned a dollar amount in the Schedule of Values based on known information at the time of budget development. As each task is further defined, its budget is adjusted to reflect the change in quantities (or scope). At regular intervals, typically just prior to the Application for Payment, the control budget is updated to acknowledge the change in quantities. Most often the updating of the budget is done with the owner, design team, and contractor all in agreement with the new quantities and values. When agreed upon, the new budget is called the *Quantity Adjusted Budget* (QAB). Here is the formula for calculating EV with this approach:

$$\text{EV} = \text{Percent Complete} \times \text{Quantity Adjusted Budget}\left(\text{QAB}\right)$$

It is not uncommon for percentages of completed work to decrease after new QABs have been introduced. While the appearance of "going backwards" is always considered disheartening, it is the nature of variable quantity work. In turn, as the quantities are adjusted up, there is typically a corresponding extension of time to the schedule to perform the additional work.

The reader is offered a word of caution in using the Variable Budget approach. Most cost-plus contracts commence work in advance of having completed documents. This is done under the moniker of fast-track delivery, which by its nature has a high degree of *rework*. Rework is defined as taking recently completed work apart and redoing it to accommodate changes in the design. Rework adds costs that are often substantial to a project, and as such it should be tracked separately. This is especially true if the reporting of costs on rework is a contractual requirement. Even though the client often pays for the rework, it would be incorrect to include the added cost for the rework in the percentage complete calculation. The result would be an erroneous value for the percent complete of the task or the project. To avoid skewing the data, cost accounts are set up for rework to be tracked independently.

Choosing a Fixed or Variable Approach

The choice of which approach to apply is often a matter decided by the contract. In other cases, it is dictated by the project conditions or by choice. Stipulated sum contracts orchestrated under the design/bid/build methodology are best served by the Fixed Budget approach. For projects that have been started with incomplete contract documents and flexible budgets, the Variable Budget approach is used. This method is most responsive to the varying quantities and maturing budgets that occur as the project becomes fully defined. Each approach has definitive characteristics that reflect the desired control.

The Fixed Budget approach is defined by the following characteristics:

- Simplified tracking and bookkeeping offer less chance for human error.
- The performance data (SPI, SV, CPI, and CV) are subject to erroneous results if the cost account budgets are not accurately distributed.
- The cost account budgets provide a fixed target that is easy to visualize and interpret.
- It provides a direct evaluation of cost and schedule performance.
- It requires a separate system for the evaluation of productivity.

The Variable Budget approach has the following characteristics:

- It requires more frequent attention to data management due to constantly changing baseline/budget values.
- Special provisions and data management are needed for tracking rework costs and time.
- It is ideal for projects with fluctuating budgets and partially defined documents.
- It can be used during initial phases of a project when there is a variable budget and then can be converted to a Fixed Budget approach when all the parameters are known.
- It provides a direct evaluation of productivity when there is no change in wage rate.
- It provides a moving budget that has a direct relationship to the actual quantities of work and the productivity rate for the tasks.
- It requires a supplementary system for the evaluation of cost performance if used with a fixed budget.

Regardless of the approach employed, measuring performance hinges on two key computations:

- Comparing earned value with the planned value
- Comparing earned value with its actual costs

These comparisons are made at the cost account level or even at the cost element level.

Schedule and Cost Performance

The concepts previously mentioned provide a process for determining the percent complete and earned value of various tasks or the entire project.

The next step is to analyze the results for determining how well the work is proceeding in accordance with the plan. The earned value system is particularly adept at providing the information with little additional calculation.

Until now the discussion has centered on planned value and earned value amounts. However, there is a third category that is needed for the analysis of performance. Two of these categories were introduced at the end of Chapter 6 Integrating the Schedule and the Budget. Consider the following outline of the three categories needed for analysis:

- Budgeted dollars (or labor-hours) to date is what the contractor had planned to do and is called the *Budgeted Cost of Work Scheduled*[1] (BCWS). This is also known as the Planned Value (PV).
- Earned dollars (or labor-hours) to date is what the contractor actually did and is called the *Budgeted Cost of Work Performed* (BCWP). This is also known as the Earned Value (EV).
- Actual dollars (or labor-hours) expended to date is what it costs the contractor and is called the *Actual Cost of Work Performed* (ACWP). It is what it costs the contractor to earn the value of the BCWP. This is also known as the Actual Cost of Work Completed (AC).

The newly introduced category of ACWP is the final component in the trinity of Earned Value Management. It is defined as all actual costs measured against the budget amounts for the cost elements in each cost account.

Performance measured against schedule is the comparison of what was actually done as compared to what was planned to have been done. In other words, it is a ratio of earned value to budgeted value. It is a measure of schedule efficiency. If budgeted dollars (or labor-hours) are less than earned dollars (or labor-hours), it translates to performing more work than planned. It also suggests that the task may be ahead of schedule. If budgeted dollars are more than earned dollars, it means less work is being performed than planned and the task is more than likely behind schedule.

Schedule performance efficiency can be measured using the *schedule performance index* (SPI):

$$\text{SPI} = \frac{\text{Earned Value}\left(\text{in dollars or labor} - \text{hours}\right)}{\text{Budgeted Value}\left(\text{in dollars or labor} - \text{hours}\right)}$$

or

$$\text{SPI} = \frac{\text{BCWP}}{\text{BCWS}}$$

For SPI values that are equal to 1.0, it means that the work is progressing exactly as planned. For values of less than 1.0, the work is behind

[1]Budgeted Cost of Work Scheduled is also known as Budgeted Cost *for* Work Scheduled. The same applies to Budgeted Cost of Work Performed. Either is acceptable.

schedule and considered an unfavorable performance. For values greater than 1.0, the work is progressing faster than planned and is recognized as a favorable performance.

It is also possible to report the difference between planned and earned value in monetary terms as a variance, instead of as a ratio. This is called the *Schedule Variance* (SV):

$$SV = \text{Earned Value} - \text{Budgeted Value}$$

or

$$SV = BCWP - BCWS$$

The SV is calculated in terms of the difference between the work actually performed and the amount of work scheduled to be completed in any given time period. It can be expressed in dollars or labor-hours. Variances that result in positive numbers (dollars or labor-hours) are considered favorable, and those that are negative are not favorable. Variances of zero, mean that the work is "as planned" for schedule.

Schedule performance reported as a variance (SV) or as a ratio (SPI) can be confusing due to the use of the word "schedule" since SV and SPI are not measured in days, the typical unit of schedule. Nonetheless, they are still measures of schedule performance.

Table 7.2 is an example of the SV and SPI calculation for a single arbitrary Task A.

Table 7.2 Calculation of SV and SPI

Cost Account: Task A				
Cost element	**BCWP**	**BCWS**	**Schedule Variance (SV)**	**Schedule Performance Index (SPI)**
Materials	$37,885.00	$36,778.00	$1,107.00	1.03
Labor	$66,890.00	$62,345.00	$4,545.00	1.07
Equipment	$3,300.00	$3,280.00	$20.00	1.01
Labor-hours	1,529	1,425	104	1.07

A quick review of the cost account reveals that Task A performed better in all cost elements than scheduled.

Performance measured against cost is the comparison of what it was planned to have cost the contractor to what it actually cost. Cost performance is measured in much the same way as schedule performance is measured by using the *Cost Performance Index* (CPI):

$$CPI = \frac{\text{Earned Value}\left(\text{in dollars or labor} - \text{hours}\right)}{\text{Actual Cost}\left(\text{in dollars or labor} - \text{hours}\right)}$$

or

$$CPI = \frac{BCWP}{ACWP}$$

For CPI values that are equal to 1.0, it means that the cost of the work is exactly as planned. For values of less than 1.0, the work is costing more than what is being earned. This is considered an unfavorable cost performance. For values greater than 1.0, the work is costing less than the amount earned and is recognized as a favorable cost performance.

Similar to schedule performance, it is also possible to report the difference between planned value and cost in monetary terms as a variance, instead of as a ratio. This is called the *Cost Variance* (CV):

$$CV = \text{Earned Value} - \text{Actual Cost}$$

or

$$CV = BCWP - ACWP$$

CV can be expressed in dollars or labor-hours. Variances that result in positive numbers (dollars or labor-hours) are considered favorable, and those that are negative are not favorable. A variance of zero means that the work is "as planned" for cost.

The reader should note that in cost performance measurement, the planned value, especially as represented in the Schedule of Values, is often burdened with overhead and profit. The planned value referenced is that which is represented by the cost account. The difference between the two is the application of overhead and profit. In the case of a Schedule of Values, which has separate individual lines for overhead and profit, the planned value (in the SOV) is the same as the planned cost and can support a direct comparison in computing the CV.

Table 7.3 is the calculation of the CV and CPI for our same arbitrary Task A.

Table 7.3 Calculation of CV and CPI

Cost Account—Task A

Cost element	BCWP	ACWP	Cost variance (CV)	Cost performance index (CPI)
Materials	$37,885.00	$33,789.00	$ 4,096.00	1.12
Labor	$66,890.00	$56,856.50	$10,033.50	1.18
Equipment	$3,300.00	$3,100.00	$200.00	1.06
Labor-hours	1,529	1,300	229	1.18

The data reveals that Task A is ahead of anticipated cost performance in all cost elements.

One final calculation used as a measure of total project performance is called the *Total Variance* (TV). It is the difference between the budgeted cost of work scheduled (BCWS) and the actual cost of work performed (ACWP).

$$TV = BCWS - ACWP$$

A positive value for the TV indicates that the project as a whole is under budget at the time of analysis. It indicates that there has been less spent by the analysis date than originally planned. It is less useful because it does not address the reason behind the underrun. It can also be the net sum of positive and negative performing tasks, without alerting the reader to the detail of the problems. There is also a formula for calculating the percent total variance (PTV):

$$PTV = \frac{BCWS - ACWP}{BCWS}$$

Again, TV and PTV are "big picture" tools and not that helpful to the project manager diagnosing the status of the project.

Summary and Key Points

Most analysis is performed by computers to save time and increase accuracy. Computers allow the project manager to do analysis on hundreds or perhaps thousands of cost accounts that provide the team with real-time data for decision making.

Assigning correct amounts to the scheduled value in a Schedule of Values is crucial. It should be the difference between cost and cost plus overhead and profit. If the values are not assigned correctly, it will provide skewed data that cannot be used to track performance correctly.

The first step in analyzing progress is measuring the amount of work performed. There are six methods for measuring the quantity of work performed. Once the quantity of work has been determined, it must be compared with the amount of work that was planned to be accomplished. This process is called Earned Value Analysis, and it is the basis for payment to the contractor. There are two approaches for performing Earned Value Analysis: the Fixed Budget and Variable Budget approaches. Each has unique characteristics that make them appropriate for certain applications.

A very important tool for the project manager is the measurement of performance, both schedule and cost. This is done by calculating the efficiency of both cost and schedule performance and is expressed in ratios as CPI and SPI, respectively. It can also be reported as a difference in cost represented by a dollar value. These are the CV and the SV, respectively.

Chapter 8 Analyzing and Reporting Variances in Schedule and Cost will explore the reasons behind variances and the information that can be derived from them.

Key points of this chapter are:

- The assigning of accurate scheduled values for tasks is the first crucial step in project control.
- Accurately measuring schedule performance, or what has been complete is the second step.
- Depending on the contract type, budgets can be fixed or variable.
- There are three key metrics in project control: BCWS, BCWP, and ACWP used to pinpoint how well a task is performing.
- The CPI, SPI, CV, and SV are all mathematical formulas for determining performance.

Chapter 7 Calculating and Analyzing Progress: Questions for Review

1. There are generally six recognized methods for calculating progress of a construction task. Identify all six.
2. Explain how the Start/Finish approach assigns percentages of completion to a task.
3. What are the disadvantages of the Experience/Opinion approach to determining percent complete?
4. The process for determining the progress on an individual task or the project as a whole is called?
5. Earned Value Analysis can only be applied to contracts with fixed budgets. True or False?
6. What is the formula for determining the earned value (EV) of a task?
7. Define Quantity Adjusted Budget and explain when it is used.
8. Explain the difference between the ACWP and the BCWP.
9. What does a Schedule Performance Index (SPI) of less than 1.00 tell the project manager?
10. Is it possible to have a SPI of less than 1.00 and a CPI of more than 1.00? What does it represent?

CHAPTER 8

ANALYZING AND REPORTING VARIANCES IN SCHEDULE AND COST

Having determined that either the schedule or the cost of the plan is not proceeding as intended, the project manager must review and analyze the supporting data. He or she must then decide if action needs to be taken or if the variance is part of the normal course of events. Chapter 7 Calculating and Analyzing Progress discussed the proper methods for measuring progress and converting those measurements to earned value. The discussion was expanded to include the efficiency ratios that are helpful in identifying any variations. The next step is to consider the scale of the variances and efficiencies and what impact they may have on the project, if any at all.

There is a reason for most variances; some are easily determined and others require a more in-depth investigation. This chapter will explore the presentation of the analyses, how they are interpreted, potential causes of the variances, and, finally, how to identify their root causes.

Understanding Project Analyses

There are numerous ways in which project data can be viewed. As noted previously, data streams from a variety of sources must be compiled and then analyzed to have any real benefit to the team. While the project manager analyzes one data source at a time, all must be considered. Without schedule and cost, the project manager can be misled. One of the methods is to present the information graphically. Some of the more common graphic views are explored in this chapter.

S-Curves

The S-curve is a graphic illustration that most project managers in the construction industry are familiar with. An S-curve is a *sigmoid function*, which is a mathematical process with results that, when plotted on an X-Y axis, vaguely represent the letter "S." S-curves are generated by software programs, but an explanation of what is entailed is still warranted.

Many processes, including the complex system that we know as the construction project, exhibit a slow buildup that accelerates to a peak

Project Control: Integrating Cost and Schedule in Construction, Second Edition. Wayne J. Del Pico.

loading of labor and equipment, followed by a tapering off and demobilization at Substantial Completion. When progress, measured in dollars or labor-hours, is plotted against time, the resulting S-curves can be used to visually demonstrate progress for a variety of tasks or the project as a whole.

As part of the control process, the project manager must monitor actual progress against planned progress for the entire project. The Earned Value S-curve allows the project manager to view the relationship between three main measurement values: BCWS, BCWP, and ACWP. It illustrates the cumulative progression of work over time for each value.

When the results of all three are plotted on a graph of dollars versus time, it is possible to determine at any given date (status date) the difference between planned, earned, and actual. It is displayed in terms of deviation from the planned schedule and cost. The EV S-curve (EVS) forms a historical record of what has happened over time (Figure 8.1). Analyses of EVSs allow the project manager to quickly identify project growth, slippage of tasks, or potential problems that could impact the project as a whole if no corrective action is taken.

It should also be noted that while most example graphs show smooth flowing lines for the actual cost of work performed (ACWP), it is rarely the case. Graphed lines for ACWP will ebb and peak with job conditions.

Understanding the S-Curve Data

The EVS provides a tremendous amount of data, some of which can be very confusing as well as informative. Before proceeding with interpreting the data, a review of some of the basic information on the EVS might be helpful.

For starters, the graph is plotted on an X-Y axis. The horizontal or X-axis is traditionally reported as a function of time. The unit of time will

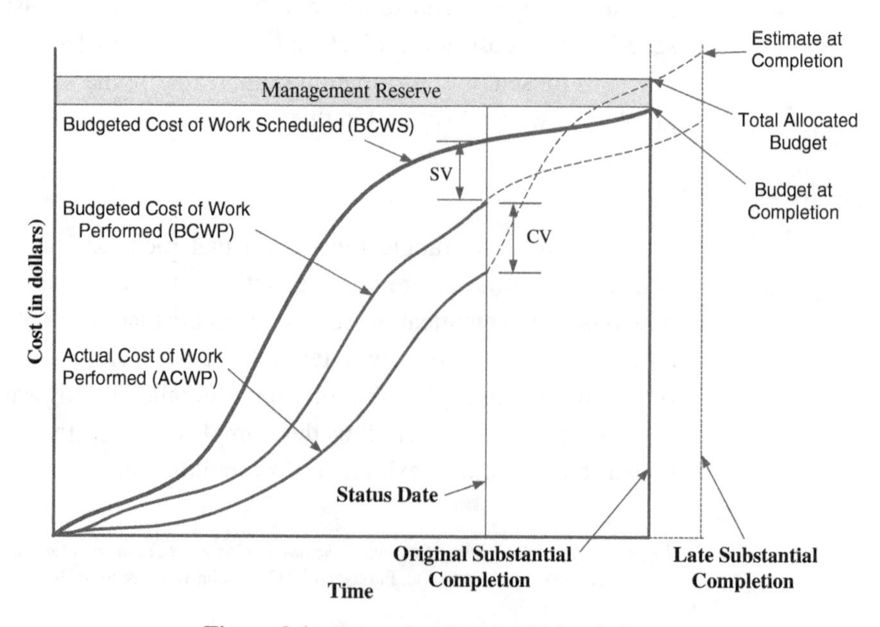

Figure 8.1 Example of Earned Value S-Curve

vary with the reporting cycle frequency, but is most often the week. It should match the reporting cycle for cost and schedule data so that the graph can be coordinated to specific data from the individual information streams. The X-axis is labeled "Time" and progresses from the commencement of work to the scheduled completion date of the project.

The vertical or Y-axis is shown on the left side of the graph. The units for the Y-axis can be dollars or labor-hours depending on the analyses required. This example will focus on dollars. The scale of the Y-axis should start at zero and finish just beyond the total budget amount. The increments should be proportional to the X-axis to maintain a uniform appearance both vertically and horizontally. Dollar amounts on the vertical scale are cumulative.

At the far right of the graph is the completion date of the project. It is shown as a vertical line parallel to the Y-axis. Standard convention is that the completion date is the Substantial Completion date, but both Substantial and Final Completion dates may be shown.

A horizontal line at the top of the graph starts at the total budget amount on the Y-axis and ends at the completion date(s). This is called the *Budget at Completion* (BAC). It represents the total anticipated expenditures on the *Performance Measurement Baseline* (PMB). The PMB is the metric established in the planning phase that will be used for comparing performance. This horizontal line is also recognized as the Original Budget and may actually be a target budget set by the project manager or senior management for the contractor. The target budget is the estimated amount less negotiated buyout obtained from subcontractors and vendors.

A second horizontal line above and parallel to the BAC represents the *Total Allocated Budget* (TAB). The TAB, as its name would imply, is the total budget available for the project. This is the budget, less the fee as represented by the as-bid estimate. For the contractor, it represents the negotiated savings between the estimate and the committed costs.

The difference between these two horizontal lines is called the *Management Reserve* (MR). MR is a contingency held internally by the contractor.[1] The MR contingency allows the project manager to fund

[1] The reader should note that Management Reserves can be established by the owner/client as a matter of course. These can be internally held or as part of special types of partnering contracts. In contrast to the MRs noted previously, these MRs are acknowledged by both the owner and the contractor. There are very specific rules for drawing funds from the MR; foremost among them is approval from the owner. MR funds are established early in the process with the help of the owner's construction manager and/or the design team and are normally a percentage of the construction budget. Renovation or unique one-of-a-kind projects have large MRs. Similar to contractor-held MRs, when the owner-funded MR is added to the Current Budget, it is called the Total Allocated Budget (TAB). The TAB is the sum of all budgets for work (or potential work) funded under the contract. It is meant to take up the overruns caused by change orders or other unforeseen latent conditions. Again, this is strictly as viewed from the owner/client perspective.

overruns without touching the original profit. Contingencies held by contractors under a stipulated sum contract are rarely, if ever, revealed to the owner and, if not used, are moved to the profit column. It should also be noted that for projects that are competitively bid with a fixed price, most contractors do not add contingencies to their bid. It is considered an unwarranted addition or nonresponsive to the bid. It is, however, fairly common to create a management reserve or contingency from the funds generated by any negotiated buyout or overhead savings from a shortened schedule. It is the difference between the estimated amount and the committed amount for a line item. This provides a reserve from which to draw without impacting the original as-bid profit.

As the budget is increased (or reduced) due to the addition of change orders, the horizontal line that represents the BAC moves vertically up (or down) the Y-axis. The MR, however, remains the same unless costs are added to the budget without a commensurate compensation from the owner. An example would be the addition of an internal change order. Internal change orders are between the general contractor and a subcontractor with no pass through to the owner.

The Budgeted Cost of Work Scheduled (BCWS) is shown as a solid line (S-shape) on the graph from commencement to completion. Since the BCWS is the performance measurement baseline developed during the planning stage, it terminates at the completion date. Logically, this is because it is assumed from the onset that the contractor will achieve delivery as specified in the contract. It would not be contractually prudent, not to mention acceptable to the owner, to show the project with a delivery date beyond the completion date specified in contract. Costs associated after the completion date can have a negative impact on the fee if there is no other source to pay for that cost.

For both the Budgeted Cost of Work Performed (BCWP) and the Actual Cost of Work Performed (ACWP), the graphed lines are solid only to the status date. Solid lines in this example indicate known entities versus unknown. The status date is the vertical line on the graph that represents the current date, or the time in which the analysis was conducted. To the right of the status date, the graphed lines are shown dashed to indicate that they are forecasted numbers based on prior performance and current trends. Values to the right of the status date are not foregone conclusions and, in many instances, can be modified by the actions of the project team.

In Figure 8.1, the ACWP is shown exceeding both the BCWP and the BCWS, which indicates that the contractor has spent more than anticipated and even more than earned. The dashed vertical line to the far right indicates that the project was delivered after the original scheduled delivery date. Many late deliveries incur liquidated damages that are not reflected in the ACWP due to the fact that they are a penalty and not a true reflection of work performed.

Some scheduling software packages generate S-curves as part of their normal reporting functions; most, however, do not. For those that don't, the reader is directed to research third party software applications that are compatible with the specific software being used. Also, most software generates the various curves in different colors for ease of identification.

Interpreting the S-Curve Data

It would be professionally unsound, if not impossible, to create absolute rules for interpreting the data that can be obtained from an EVS. Instead, it is the author's preference to provide some guidelines for consideration when interpreting the data. In any event, changes that deviate substantially from the BCWS warrant the investigation of the project management team to determine a root cause.

Here are a few generalizations that may offer guidance in interpreting the data:

- The line for ACWP ideally should be below the line for BCWP on the graph. This represents that cost to perform the work is less than what is earned, indicating a potential profit.
- The completion line should move to the right as extensions of time are granted to the contract.
- If the project experiences actual acceleration directed by the owner/client, the completion line will move to the left. This may also increase the BAC if there are costs associated with actual acceleration.
- Actual acceleration will change the BCWS from the date the acceleration is enacted to the new completion date. The curve will become more vertical.
- BCWS lines with sharp rises over short time periods (horizontal runs) can be an indicator of aggressive, potentially unrealistic scheduling.
- BCWP lines that consistently exceed (are above) BCWS are a reasonable indicator of work progressing faster than anticipated, and may also indicate that the BCWS was insufficiently aggressive.
- Large initial separations between the lines for ACWP and BCWP with BCWP above the ACWP may be an indicator of excessive front loading. The positions of the two lines will reverse near the end of the project to reflect the shortfall.
- It is not uncommon for the ACWP graph to intersect and exceed the BCWP at certain times during the project. Sharp rises in the costs of a task may be an indicator of a problem or a negative trend. It can also be the routine posting of a large invoice for materials or equipment.

- A sharp rise in the ACWP line could be attributable to an acceleration of work or an increase in crew size. This can often be expected after the resolution of issues that have stymied progress. It may reflect overtime costs.
- Delays are often represented by steep dips in the line for the ACWP, which may be mirrored by the graph for the BCWP.
- In an ideal world where everything went according to plan, the BCWP would be graphed directly over the BCWS, and the ACWP would be parallel to the BCWP but below and to the right of it.

Tracking Gantt Chart

Clearly one of the best sources of data is the *Tracking Gantt* chart. Tracking Gantt charts compare actual start and finish times to baselines for individual tasks. Professional scheduling software is equipped to set a baseline schedule and, with the appropriate data entry, monitor how closely the actual schedule performs in comparison to the baseline. In an easy-to-follow visual format, the baseline for the task is identified as a bar with start and finish dates defined on the timescale. Updates to the baseline are shown in bar style beneath the task with actual start and finish dates. The different bars are illustrated in contrasting colors or infill patterns to avoid confusion.

When the critical path is defined by linking the tasks, any deviation from the baseline illustrates the impact on the overall schedule. In the case of delay to the critical path, the project manager is alerted to the need for corrective action, or at the least investigation as to the cause. Very few tools in the project control toolbox have the immediate visual impact of the Tracking Gantt.

Activities in process show the progress to date and show the remaining scheduled duration, assuming the task will complete on time. Activities that have not yet started show a revised start date based on links to their predecessors. Multiple updates of the Tracking Gantt can often show trends for a task and possibly identify the cause of the delay. Figure 8.2 is a fragment of a schedule from Microsoft Project. Tracking Gantt charts, to a lesser degree of accuracy, can be used to monitor costs as well. Many scheduling software products have the ability to produce reports in addition to the graphic schedules. There are Cash Flow and Earned Value reports to name two. Updated correctly, the reports can show a wealth of information that can be used in the decision making for corrective action.

Control Charts

The control chart is another management tool to monitor schedule performance (Figure 8.3). It tracks past and current performance by update and can be used to predict future performance with modest accuracy. It is

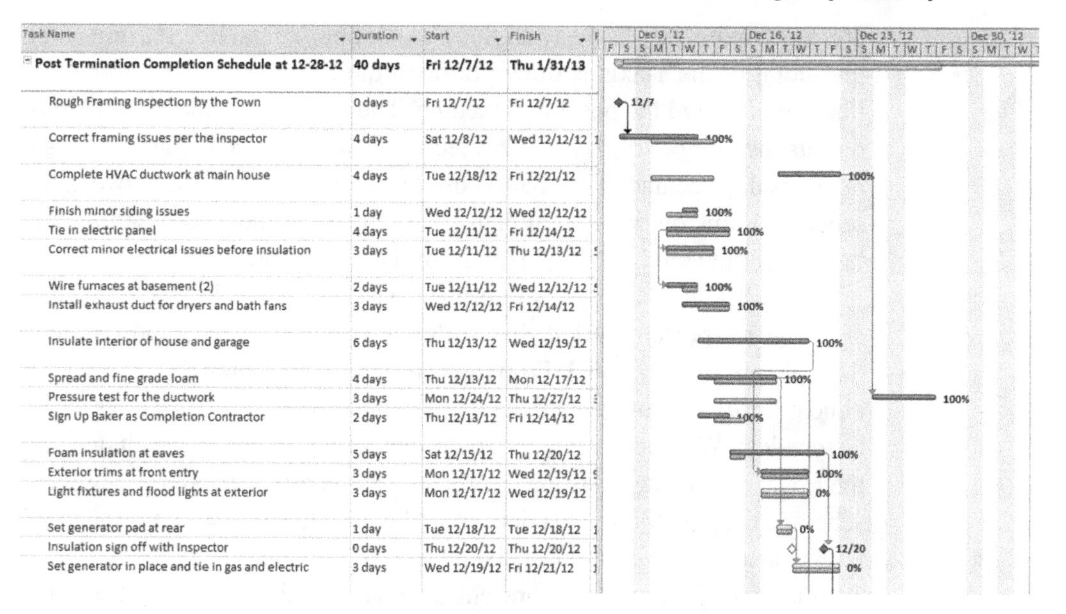

Task Name	Duration	Start	Finish
Post Termination Completion Schedule at 12-28-12	**40 days**	**Fri 12/7/12**	**Thu 1/31/13**
Rough Framing Inspection by the Town	0 days	Fri 12/7/12	Fri 12/7/12
Correct framing issues per the inspector	4 days	Sat 12/8/12	Wed 12/12/12
Complete HVAC ductwork at main house	4 days	Tue 12/18/12	Fri 12/21/12
Finish minor siding issues	1 day	Wed 12/12/12	Wed 12/12/12
Tie in electric panel	4 days	Tue 12/11/12	Fri 12/14/12
Correct minor electrical issues before insulation	3 days	Tue 12/11/12	Thu 12/13/12
Wire furnaces at basement (2)	2 days	Tue 12/11/12	Wed 12/12/12
Install exhaust duct for dryers and bath fans	3 days	Wed 12/12/12	Fri 12/14/12
Insulate interior of house and garage	6 days	Thu 12/13/12	Wed 12/19/12
Spread and fine grade loam	4 days	Thu 12/13/12	Mon 12/17/12
Pressure test for the ductwork	3 days	Mon 12/24/12	Thu 12/27/12
Sign Up Baker as Completion Contractor	2 days	Thu 12/13/12	Fri 12/14/12
Foam insulation at eaves	5 days	Sat 12/15/12	Thu 12/20/12
Exterior trims at front entry	3 days	Mon 12/17/12	Wed 12/19/12
Light fixtures and flood lights at exterior	3 days	Mon 12/17/12	Wed 12/19/12
Set generator pad at rear	1 day	Tue 12/18/12	Tue 12/18/12
Insulation sign off with inspector	0 days	Thu 12/20/12	Thu 12/20/12
Set generator in place and tie in gas and electric	3 days	Wed 12/19/12	Fri 12/21/12

Figure 8.2 Schedule Fragment from MS Project

a summary view that focuses on the project's critical path in contrast to individual tasks.

A study of Figure 8.3 indicates that the data is plotted against an X-Y axis. The horizontal or X-axis starts with project commencement at zero and is delineated by the number of expected update reports. The vertical or Y-axis is divided into two horizontal sections with zero again as the starting

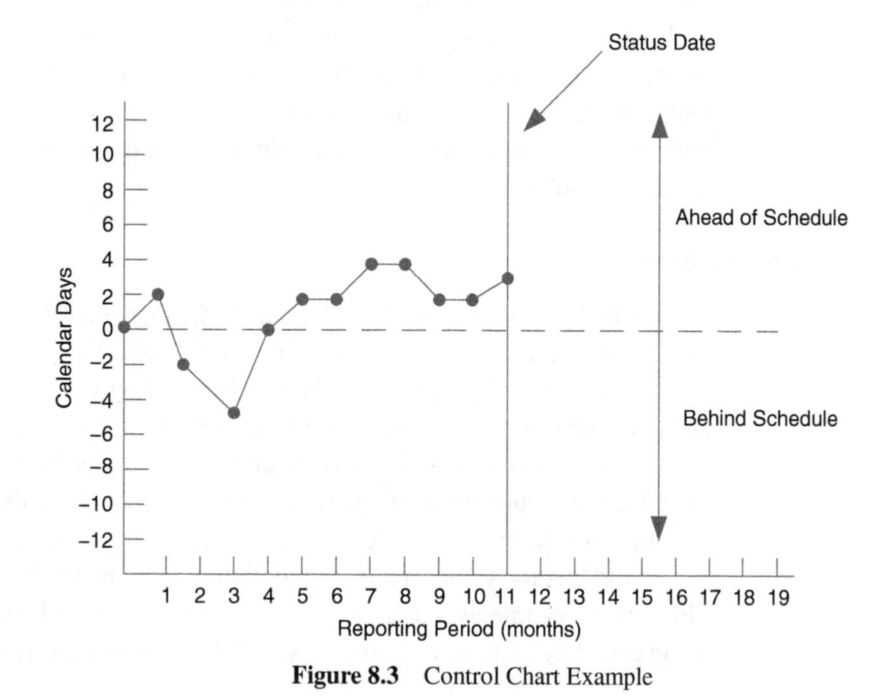

Figure 8.3 Control Chart Example

point approximately halfway up the Y-axis. A horizontal line starting at the mid-point of the Y-axis is drawn through the last update. This horizontal line is considered the baseline schedule. The scale above the baseline represents the number of days the project is ahead of schedule, and the scale below is days behind. The chart is used to plot the difference between the scheduled time on the critical path (represented by the zero baseline) and the actual time on the critical path as of the report date.

Figure 8.3 indicates that the project was two days ahead of schedule at the first reporting period, but suffered a setback that would continue through reporting period 3. From reporting periods 3 through 6, there is an almost steady improvement that results in once again being two days ahead of schedule. Without knowledge to the contrary, it might be concluded that the improvement suggests that corrective action after period 3 brought the schedule back on track. This data could be used to suggest that if the trend is sustained, the project will deliver ahead of schedule.

Since most task times represent average durations, four updates that each support a trend usually indicate, with high probability, that there is a reason for the improvement. The project manager should attempt to ascertain the cause and potentially exploit it to maintain the trend.

The disadvantage to the control chart is its lack of detail. It is a summary tool with its greatest appeal on projects of long duration. From the update points alone, there is insufficient data to determine anything other than a positive or negative position at that date. Control charts are most often used to track progress toward a milestone or as an update to upper management. Slippage that results in work a few days behind schedule in the first half of the project is less alarming than if it occurs in the last quarter of the schedule. Therefore, control charts have the advantage of representing the time remaining for correcting any problems. As remaining time in the schedule expires, the flexibility and opportunity to rectify the delay diminish.

SPI and CPI Charts

Similar to the control chart, the SPI and the CPI can be plotted on an X-Y axis graph to track each over time (Figures 8.4 and 8.5).

In each of the charts, the X-axis is delineated by the reporting period and the Y-axis is divided into two horizontal sections with 1.00 as the starting point approximately halfway up the Y-axis. For both the SPI and the CPI, 1.00 indicates an as-planned performance. Above the horizontal line are delineations that increase in .10 increments: 1.10, 1.20, and 1.30. Below the line the values are decreased in .10 increments. As the CPI and SPI values become available at each reporting period and are plotted on the graph, they indicate a history of schedule and cost performance for the

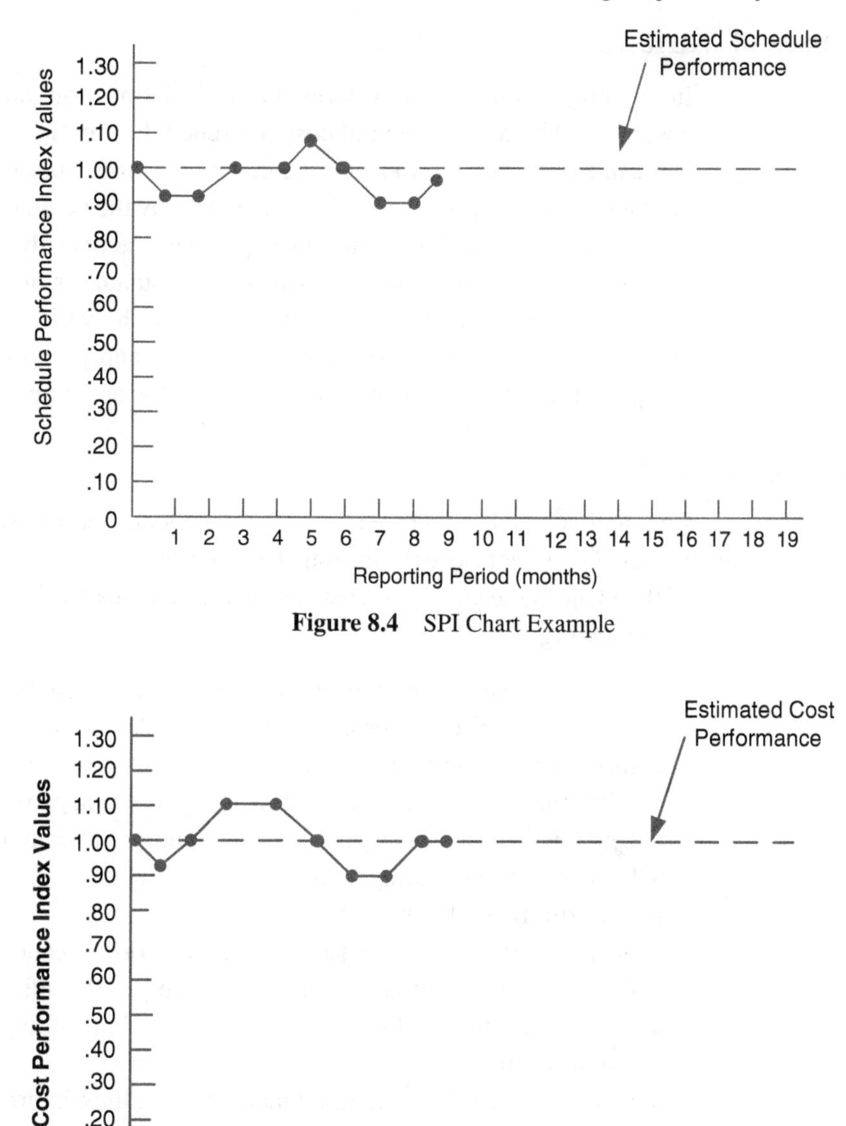

Figure 8.4 SPI Chart Example

Figure 8.5 CPI Chart Example

project. When the variance threshold for SPI and CPI (discussed in the section to follow) are highlighted in red, it is a great visual tool for indicating when the project or one of its tasks has a potential problem.

While both charts are great tools for summarizing performance, they do not provide the detail as to the cause of the problem. They are historical in nature and not always a real-time alarm to the problem causing the variance.

Causes of Variances

In any project a *variance* is defined as a deviation from any schedule or cost value. The reasons behind cost and schedule variations are numerous and can be attributable to a myriad of reasons from poor estimating and scheduling practices to gross mismanagement by the contractor. Assuming that estimating, scheduling, and management practices by the contractor are commensurate with professional practice standards, it would be prudent to consider some other potential causes of the variances. A standard approach is to review the simplest causes first and progress to the more complex. It is amazing the number of causes that are simple fixes.

Cost Variances

Cost and schedule variances are interdependent, yet a positive schedule variance does not necessarily translate to a positive cost variance. Some of the identifiable causes of cost variances, both positive and negative, are the following:

- Poorly defined scope during the buyout phase that can often add to subcontractor/vendor costs as the work packages are executed
- Incorrect distribution of dollars to cost accounts in the budgeting phase that may show up as excessive positive or negative cost variance
- Unrealistic or overly aggressive target budgets at completion (BAC)
- Excessive front loading of costs
- Performing work out of sequence
- Scope creep due to the rejection of change order requests
- Excessive use of premium wages with little or no benefit to the schedule
- Crew sizes too small or too large for the task, resulting in inefficient use of resources
- Hastily or partially purchased materials resulting in premium costs
- Frequent delays of crews from inadequate material stores on-site
- Insufficient/incorrect tools or equipment for the task
- Excessive handling, freight, and storage costs
- Changes in labor rates, benefits, or labor burden from estimated values

Schedule Variances

Some of the more common causes for schedule variance include:

- Continual or excessive interruptions to production forces
- Lack of adequate and clear direction to workforce
- Rework due to inadequate supervision or poor direction
- Provision of material stores not adequate to prevent interruption in production

- Use of overtime when there is no tangible benefit to the schedule
- Unforeseen latent conditions resulting in a work stoppage
- Improper equipment or tools for the task
- Crew size not appropriate for the task, typically undersized
- Delays due to equipment breakdown
- Performing work out of sequence with reduced performance
- Late start of a critical task
- Incorrect linkage of a critical task to a noncritical task
- Loss of continuity due to change in supervision
- Change in crew composition (tradesmen vs. apprentices)
- Attrition of labor force
- Understaffed subcontractor labor workforce
- Scope more extensive than was initially planned
- Staging area too far from work area
- General malaise, indifference, or fatigue of the workforce

Variance Thresholds

Monitoring project success by tracking cost and schedule variances becomes a little more realistic and considerably easier if all variances do not require immediate investigation. Many variances to both schedule and cost are a result of common everyday issues like timing. Variances due to timing usually sort themselves out in subsequent reports. While variances are important, they are only helpful when the correct frame of reference is applied. For example, a cost variance of $1,000 warrants concern if the cost account is $10,000; however, it is of less concern to a cost account with a budget of $100,000.

In order to manage more efficiently, project managers with the help of their project teams set a *threshold* on both cost and schedule variance. A threshold is a limit or boundary of a variance expressed as a percentage of its performance index. Thresholds allow the team to know when to take corrective action and when to apply a "wait-and-see" strategy.

Thresholds are typically applied to the performance indices of CPI and SPI (discussed in Chapter 7 Calculating and Analyzing Progress). Performance indices are indicators that are relative to the size of the project and offer more useful information than variances especially in terms of setting thresholds.

Consider an example of a task with the given information at reporting period 4:

$$BCWP = \$50,000.$$

$$ACWP = \$46,700.$$

$$BCWS = \$55,000.$$

Here are the calculations of SV, CV, SPI, and CPI for the example:

$$SV = BCWP - BCWS = \$50,000 - \$55,000 = -\$5,000$$

$$CV = BCWP - ACWP = \$50,000 - \$46,700 = +\$3,300$$

$$SPI = BCWP \div BCWS = \$50,000 \div \$55,000 = .90\,(90\%)$$

$$CPI = BCWP \div ACWP = \$50,000 \div \$46,700 = 1.07\,(107\%)$$

A review of the data indicates that the project is behind schedule by 10 percent but we are ahead financially by 7 percent. If in the planning stage the team defined the thresholds for SPI at plus or minus 5 points (5 percent) and CPI at plus or minus 8 points (8 percent), it would trigger an investigation of the schedule overrun. However, the cost underrun does not warrant concern, as it is within the limits prescribed by the threshold. The SPI indicates to the project manager that for every dollar that the team planned on spending, they are getting 90 cents worth of performance. The CPI indicates to the team that for each dollar spent, the team is achieving $1.07 worth of performance. Clearly, with these thresholds in place, some investigation of the schedule slippage is required.

The difficult task is choosing the correct threshold for a task. The tasks requiring the greatest control are self-performed tasks, on the critical path. There is no uniform methodology for deciding variance thresholds. It is often the result of a discussion between team members and senior management as to which tolerances are acceptable. Thresholds may be dependent on a variety of factors such as:

- Overall project duration
- Shorter duration tasks with higher value may require tighter variances
- Contractor's experience with the task, lack of experience may require tighter variances
- Dollar value of the task as a percentage of the total contract
- Type of contract: fixed price or cost-plus
- Profit margin of the contract
- Phase of the project
- Contractor's tolerance for risk

Thresholds may be different from project to project or from task to task. Thresholds may even be permitted to change over the life of the project with greater threshold variances at the beginning than at the end of the project. The theory is that risk diminishes as the task gets closer to completion due to proficiency in the task. Again, if the actual condition revealed by tracking shows a lack of proficiency, then that condition should take precedence.

Variance thresholds are based on risk tolerance of the company for that individual project. They are rarely, if ever, driven by a single factor. They are most often a combination of multiple factors. There are some general guidelines that may be helpful:

- Variance thresholds set at less than 5 percent suggest the contractor has a low tolerance for risk.
- Variance thresholds set at between 5 percent and 10 percent suggest the contractor has a moderate tolerance for risk.
- Variance thresholds set at 10 percent or over suggest the contractor has a high tolerance for risk.

Variance thresholds for critical path tasks with SPIs and CPIs can be interpreted as follows:

- Less than 5 percent is viewed as an early warning of a potential problem.
- Between 5 percent and 10 percent is cause for investigation and possibly action.
- Over 10 percent is cause for immediate and possibly substantial action.

Higher percentages on thresholds can also be translated into taking action later, rather than earlier by the management team.

Root Cause Analysis

Once the data has been reviewed and analyzed and the problem task(s) are known, the next step is to identify what is driving the variance. In much the same manner as a doctor must diagnose and treat the cause of an illness to prevent the symptom from reappearing, the project manager must look deeper to determine the reason for the variance and then correct it to either stop further harmful effects or prevent it from reoccurring in the future.

If the project manager is unable to determine why a particular variance occurred, how will the team be able to specify corrective measures with any confidence or success? Understanding what led to a particular cost or schedule failure is the key to developing an effective and efficient plan of corrective action. The more specific the diagnosis, the more focused the remedy.

While not developed specifically for the construction industry, there is a process aimed at identifying the source of the problem event that precipitated the variance. It is called *Root Cause Analysis* (RCA). RCA is a systematic approach for discovering the factors that resulted in the failure. It identifies what actions, inactions, individuals, conditions, or processes need to be changed to prevent continued or future failures (Figure 8.6). RCA is more reactive than proactive in nature; however, it can be used to forecast problems before they happen. Effective root cause analysis

Medical Office Building, Boston, MA 02012		Project # 12-2201	Event Analysis Date: 12-12-22	
Event Description: *Drywall finish compound froze and had to be removed and re-applied causing a delay in the critical path of the schedule.*				
Ref No.	**Causal Factor Description**	**Analysis Results**	**Recommendations**	**Comments**
Causal Factor #1	Permanent heat system not available for use.	• HVAC Foreman not aware of system start-up date. • Gas not inspected and available for system use. • System start up and Test not scheduled.	• New HVAC foreman contributed to loss of communication. • Deadlines communicated better by GC super. • Confirm manufacturer start and test dates earlier.	More frequent progress checks. Poor communication between GC super and HVAC foreman.
Causal Factor #2	Temporary heaters inadequate for area heated.	• No back up plan formulated if permanent heat unusable. • Too few heaters available for rent. • Available units too small for large area to be heated.	• Have back up plan in place for "what-if" scenarios. • Find additional sources of temp heaters. • Reduce areas to be heated with temporary partitions.	Should have had a contingency plan given prior experience with gas utility company.
Causal Factor #3	Inadequate crew size to meet scheduled finish dates.	• Lack of heat forced reduction in crew size. • Heated area too small for larger crew. • Crew not asked to work overtime	• Schedule overtime. • Add a second shift. • Reduce areas to be heated with temporary partitions.	Crew should have been asked to work overtime or a second provided.
Causal Factor #4	Gas utility meter on the building not installed on schedule.	• Plumbing foreman late in requesting inspection. • Gas not inspected in sufficient time. • Gas meter installed late.	• Plumbing foreman provide longer lead time for inspection. • Allow more time for local inspector to inspect. • Local gas utility need minimum 15-days after inspection sign off.	Poor management was a contributing factor in not coordinating utility sooner.

Figure 8.6 Root Cause Analysis Matrix

over time can be used to structure opportunities for improvement in the construction process. If, for example, repeated cost and schedule overruns in a normally repetitive process can be sourced to an antiquated piece of equipment, RCA could provide the data for making a business case to replace the faulty piece of equipment.

Root causes can generally be defined as follows:

- The specific underlying cause of a problem
- Within the purview of the project team to correct
- Causes that can be identified and isolated
- Problems for which practical effective solutions can be found

There are four major steps to the RCA process:

Step 1—Collect the data required to fully identify the problem. While data from the S-curve, CPI, and SPI may be the trigger to the investigation, additional who, what, where, and when questions are needed. Make assumptions only if necessary.

Step 2—Organize the data into factors that were contributors to the event. Create a written list of all contributing factors. The factors can be laid out sequentially or chronologically leading up to the problem. This particular step allows the project team to see the interrelationship between the factors.

Step 3—Identify the root cause of the event. Find the base reasons for each contributing factor in order to understand how to correct the failure. Verify that assumptions made are valid.

Step 4—The final step in the process is to generate a recommended solution and craft a plan to implement it. Once implemented, it must be tracked and analyzed to ensure it is the correct solution. Frequently, adjustments to the initial plan are needed to optimize the fix.

As with any acknowledged problem, those accountable for solving the problem want everyone to know when it's corrected. For any major RCA, a dedicated report is generated. However, for most routine corrections, a redacted version of the events is adequate in the normal course of update reports.

Reporting Project Status

Much of the data that is analyzed comes from multiple sources. They include daily reports, job cost reports, look-back schedules, updated tracking schedules, and a myriad of other, less formal means. Some of the data may even be contradictory and require further analysis. Regardless of the source, it all is part of the snapshot that is the status of the project.

It is the responsibility of the project manager to sift through the data and decide which issues are of importance and what doesn't make the cut, then share it with the team and other stakeholders. The simplest approach that conveys the information is always the best and most effective communication tool.

Effective communication is the key to successful project management. Sharing the correct information with the right people in a timely manner is paramount to project success. The status report is a way of ensuring that all team members are on the same page and have the latest progress update information. The project status report is a regular, formalized report on project progress as compared with the project plan.

Status reports are not meant to be creative writing assignments; they are meant to convey information to busy people. As a result, there are some key components that all status reports share:

- Reports should be concise and clearly written.
- Reports must have basic information: date, project, recipient list, how the report is delivered, etc.
- Regular status reports should follow the same format.
- Reports must convey what has been completed.
- Reports must identify what tasks are behind or in danger of falling behind.
- Reports should identify the reason for a slipped task, if known.
- Reports should identify milestones that have been achieved and those that have slipped.
- Reports should identify ongoing or potential problem tasks or issues that are being monitored.
- Reports should be written in the third person.
- Reports are less narrative and more declarative; bulleted information is acceptable.
- Reports should present the facts and avoid presenting opinion.
- Reports should remain positive, but avoid the Pollyanna syndrome (robust optimism).
- If the report expects action from a team member, it must make sure that the action is clear to the reader.
- Reports must maintain simple, but effective status reporting.
- The report must be available at the same day and time so that participants can expect it with regularity, especially if the remedy is extended over multiple reporting periods.

Good, consistent project status reporting helps guard against unexpected surprises to the participants. It is intended to provide team members with a clear view of the project at that date. It should address topics such as schedule, cost, resources, changes, disputes, and any issue requiring the focus of the team. Remember the status report is a documented history of the project and the decision making of its managers. It is different from meeting minutes since it originates with the project manager. It will become part of the project record and, as such, should be written with sufficient detail so that minimally involved individuals can understand the report at a later date.

Frequency of reporting will vary from project to project depending on factors such as project size, expectations of information, distribution by management, and project complexity with the associated risks. The most common reporting period is the week.

Project status is typically reported by the project manager to those in upper-level management, such as senior and project executives. It is also shared with functional managers, such as team leads, superintendents, foremen, and other parties needing updated information. The finalized project status report should be distributed among team members in accordance with the project's communication plan (Figure 8.7).

It is important to realize that the status report is not intended as a substitution for ongoing project communication between team members, the project manager, the client, or other parties requiring updated information to perform their duties. Status reports are simply a formal document summarizing progress since the last status report. Also remember that the recipients are busy people so avoid the *War & Peace* reports that take hours to read and comprehend. Get to the point(s)!

Summary and Key Points

Sources of information for project control are numerous and as such require careful analysis in order to be of value. One tool for analyzing and comparing data is the Earned Value S-curve (EVS). The EVS allows the team to compare planned versus actual schedule and cost data and derive efficiency factors from the data. The EVS is generated by data input into software capable of tracking tasks on a schedule with cost control. Efficiency factors can signal a problem that may require further analysis to determine the cause. While there are many causes of variances, there are some common threads to both cost and schedule variance. Some can be identified and corrected easily when they exceed the threshold set by the team. Others require a more in-depth analysis through a process called a Root Cause Analysis.

The Tracking Gantt chart can compare actual start/finish dates and durations of individual tasks and their impact on the critical path. Other types of graphic depictions such as control, SPI, and CPI charts indicate a bigger picture tracked over time.

Medical Office Building, Boston, MA 02012		Project # 12-2201	Update: 12-14-22	Last Update: 12-1-22
Report authored by Bob Smith, PM 12-14-22 5:00PM				
Electronic distribution list on last page.				
Ref No.	**General Progress Update**	**Date Completed**	**Action by**	**Comments**
1	Wood frame inspection signed off by Building Inspector Joe King. Proceeding with insulation. Scheduled insulation inspection.	12/5/2022	Bob Smith, PM	Updated schedule on website for completed milestone. This inspection was one day late due to inspector's workload.
2	Acme Insulation Contractors started insulation at main structure west end. Acme took 4-days to complete.	12/9/2022	Bob Smith, PM	Update schedule
3	There were a few minor issues. Emails identifying those issues have been sent to the appropriate subcontractors for correction before or by Thursday.	12/6/2022	Jack Jones, Super	Confirmed board delivery and crew to hang drywall.
4	Insulation signed off by building inspector. Schedule board to be delivered.	12/11/2022	Jack Jones, Super	Board scheduled for delivery 12-12-22.
5	Drywall delivery arrived mid-day and loaded by end of business.	12/12/2022	Best Drywall Co.	All loaded. No damage to structure.
6	Board hangers arrived to start hanging drywall.	12/13/2022	Jack Jones, Super	Four person crew. Will require more hangers once underway.
7	Loam had been stockpiled on the site for spreading as soon as the rain abates and the ground dries.	12/10/2022	Jack Jones, Super	Schedule for early next week if dry weather.
8	The minor issues with the siding were being taken care of today.	12/11/2022	Jack Jones, Super	Record item- no action required.
9	The front entry columns and beams have been completed.	12/10/2022	Jack Jones, Super	Record item- no action required.
10	Two half column details at the front entry were omitted. It is recommended that we do not take the siding down to install them. Architect and owner were ok form an aesthetic point of view.	12/9/2022	Bob Smith, PM and Jack Jones Super	Should be updated in the project record and on the as-builts that columns have been deleted.

Figure 8.7 Sample Status Report

As with all project information, good or bad, it must be shared so that stakeholders know the status and the project team can correct or eliminate unacceptable conditions. Effective communication through status reporting is essential to good project management.

Chapter 9 Recognizing Trends and Forecasting Performance will discuss how the data can be used to spot trends and predict future performance.

Key points of this chapter are:

- Earned value management has three key data points: BCWS, BCWP, and ACWP to address status.
- EV management can be plotted on an S-curve for a graphic view of performance.
- Measurement of all three data points must be tracked and reported regularly to determine trends.
- Schedule and cost efficiency can be different for each task with one being positive and the other negative.
- Variances should have tolerance thresholds set by the PM team or senior management for an individual task.
- Efficiency indices such as the CPI and SPI are a good starting point for determining the problem.
- Start with the simplest cause first and proceed to the more complex ones.
- Reporting project status should be factual and brief, convey only what is required for an understanding of the status.

Chapter 8 Analyzing and Reporting Variances in Schedule and Cost: Questions for Review

1. The Earned Value S-curve graphs the dollar value of three separate parameters as a function of time. What are those parameters?
2. The Budget at Completion (BAC) never changes from the beginning of the project. True or False?
3. The Budget at Completion (BAC) plus the Management Reserve (MR) is called?
4. Explain why the BAC ends at Substantial Completion on the S-curve.
5. The completion line on the Earned Value S-curve should move to the right as extensions of time are granted to the contractor. True or False?
6. Large initial separations on the EV S-curve between the lines for ACWP and BCWP with BCWP above the ACWP may be an indicator of excessive front loading. True or False?

7. Earned Value S-curves can be produced for a single task or for the entire project. True or False?
8. Explain the purpose of a Control Chart.
9. SPI and CPI charts are used to graph the respective efficiency factors so that they can be tracked over time. True or False?
10. Variance thresholds are a measure of tolerance for SPI and CPI efficiencies. Explain.

CHAPTER 9

RECOGNIZING TRENDS AND FORECASTING PERFORMANCE

One of the biggest advantages to project control is not only being able to look at past and current performance but also to try to predict what will happen at crucial points in the future of the project up to its completion. In order to do that effectively, the data introduced in Chapter 8 Analyzing and Reporting Variances in Schedule and Cost must be examined through multiple reporting periods to look for repetitive patterns in the performance. It is these patterns that are used to predict future performance.

Recognizing Trends

It is very common for patterns in the performance of tasks to develop as work progresses and the learning curve is mastered. It is especially true for labor-intensive, long-duration tasks. Repeat patterns of similar performance in construction are called *trends*. Comparing the variances of current to past reporting periods is called *trend analysis*. Trend analysis is the first step in using the tracking data to predict future performance.

Acknowledging and monitoring trends are not just about forecasting performance, but also recognizing future needs of labor, materials, and equipment for the project going forward. It allows managers to be proactive versus reactive. Trends develop over time and, as such, there is an expectation that if all things remain the same without external input, the performance trend will continue. This is a generally accepted premise that for the most part holds true. Experience has shown that if a project is hemorrhaging cash, it will continue to do so if no action is taken. However, in many circumstances, with the proper diagnosis and the right corrective action, poor performance from both a cost and schedule perspective can be arrested and brought back in line with baseline values. (Assuming that budgets and schedules were developed within recognized professional standards.)

Trend analysis is a "big picture" project management tool. While it illustrates past and current performance overall for the project or a task, it does not tell the project manager why. Trend analysis is a mathematical

Project Control: Integrating Cost and Schedule in Construction, Second Edition. Wayne J. Del Pico.

technique that uses historical results to predict future outcome. This is achieved by tracking variances in cost and schedule performance.

Trend analysis is based on a simple premise: Past performance is a good indicator of future performance. Again, this is predicated on no intervening action by the project team. Trend analysis is often considered part of variance analysis, but in the construction universe, variance analysis is immediate and trend analysis is more long-term. Variance analysis can also be more granular in the results. For that reason, the author has chosen to address trend analysis as a separate task.

As with any use of statistical data to forecast, it is limited by the quantity and accuracy of the data. Accuracy of the data is imperative to forecast any performance with even the slightest integrity. Without it, the forecast is a guess, and an unsubstantiated one at that. By quantity, we are referring to the quantity of update or tracking periods that bolster the forecast. Clearly, one can understand that forecasting the cost to complete using 3 reporting periods' worth of data is less reliable than with 23.

Every industry has its rules for the point in the schedule where sufficient data has been collected for forecasting. In other words, how much of the task/project has to be completed before the performance data collected can be used to accurately forecast future performance? In circumstances with rigid tolerances and quality control standards, such as manufacturing, that number can be as low as 10 percent. For research and development (R&D) applications, it can be as high as 50 percent. Construction is no different. However, since reliability of the forecast is the priority, there are four considerations that trigger when data becomes an effective basis for forecasting:

- Any learning period for the task has expired.
- The crew executing the work is fixed.
- The production rate for the task has stabilized.
- A minimum of three reporting periods of data has been collected.

In general, the higher the percentage of completion, the more accurate the forecast will be. Predicting the cost to complete when 90 percent of the task has been executed has a far greater chance of accuracy than when the task is 40 percent complete. The other half of that statement is with 90 percent complete, there is little left of the task to make changes, whereas at 40 percent complete, there is sufficient time to enact some changes. That said, the project manager should remember that all forecasts are not inevitable conclusions at any percent complete. Many project managers have turned their attention away from a high performing task only to revisit it later and find that task performance has plummeted for no apparent reason. Forecasts are predictions based on factual data; however, they are still predictions and they need to be watched and verified.

The control and efficiency charts (SPI and CPI) introduced in Chapter 8 Analyzing and Reporting Variances in Schedule and Cost are often used to identify trends and forecast performance. All three are examples of *trend charts*. Trend charts are used to illustrate any tracked performance data as a function of time. Their sole purpose is to increase understanding of actual performance as compared with planned performance. Regardless of the performance being measured, trend charts have some shared characteristics:

- Clear titles identifying what trend is being measured
- Dates when submitted
- Labels on the X-axis identifying the time intervals
- Labels on the Y-axis describing the subject trend
- Cumulative scales on the Y-axis with incremental grading to show variations
- Legends to define each line
- Separate lines for actual data and planned data to avoid confusion
- Explanations of any major deviations from the plan
- Vertical lines showing the status date

Frequent problems include adding too many lines for comparison. Generally, only two lines are plotted: the actual data and the planned performance of the subject trend (Figure 9.1). Another common flaw is to use too large a scale for the vertical increments. The end result is that the variations are minimal. Remember, graphic illustrations are intended to have a visual impact on the viewer. Too narrow a frame of reference will make even large variations appear negligible.

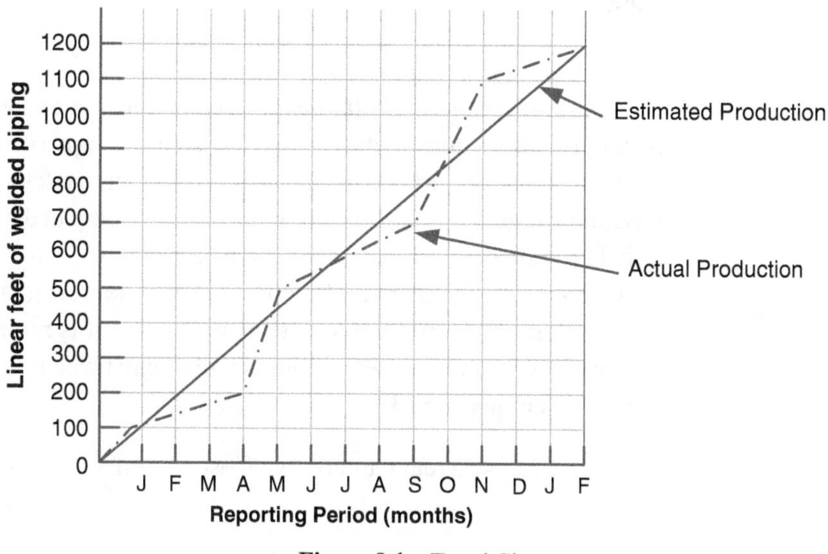

Figure 9.1 Trend Chart

Common vertical performance metrics include percent complete, labor-hours, units completed, and dollars earned.

Using Trends to Forecast

One tool that is useful in forecasting performance is the Earned Value S-curve (EVS) introduced previously. However, before forecasted data can be plotted on the S-curve, it has to be calculated. As part of the Earned Value Management system, there is a group of formulas that are used to forecast future cost in a variety of ways. These formulas help calculate four main forecasts:

- The total amount of money (or labor-hours) that will be expended at completion of the task/project if the trends continue
- The remaining amount of money (or labor-hours) needed to complete the task/project based on actual cost (or labor-hours) expended to date
- The variance at the completion of the project if the forecast is accurate
- What performance will be required to bring the task/project back on track with planned performance

Estimate at Completion

One of the most sought-after forecasted values is the total anticipated cost of the project when it is complete. This is called the *Estimate at Completion* (EAC). It is also called the *Latest Revised Estimate* (LRE). For the purposes of our discussion we will use EAC. Formulas for calculating the EAC range from simplistic to more complex. The following is the simplest version where BAC is budget at completion and CPI is cost performance index:

$$\text{EAC} = \text{BAC} \div \text{CPI}$$

This simple version is often used in early calculations when a sufficient history of data is unavailable for a more accurate approximation. The BAC is the total anticipated cost of the task (or project) as determined when the baseline budget was established. It considers anticipated cost performance only based on the CPI, and ignores any actual cost performance history.

A more complex, and often more accurate formula, considers the amount that has been spent to date as well as what is remaining. It factors in only the CPI and ignores any impact that may be caused by the schedule performance index (SPI):

$$\text{EAC}_{\text{CPI}} = \text{Actual Cumulative Cost} + (\text{Remaining Work} \div \text{CPI})$$

or

$$\text{EAC}_{\text{CPI}} = \text{ACWP}_{\text{c}} + \left[\left(\text{BAC} - \text{BCWP}_{\text{c}}\right) \div \text{CPI}_{\text{a}}\right]$$

where

$ACWP_c$ = cumulative actual cost of work performed

BAC = budget at completion from the baseline

$BCWP_c$ = cumulative budgeted cost of work performed

CPI_a = average cost performance index

The last and most definitive approach (some would characterize it as an overly pessimistic approach) considers the impact of the SPI, which is excluded in both prior methods. It is called the EAC composite:

$$EAC_{com} = ACWP_c + \left[(BAC - BCWP_c) \div (CPI_a \times SPI_a)\right]$$

where

$ACWP_c$ = cumulative actual cost of work performed

BAC = budget at completion from the baseline

$BCWP_c$ = cumulative budgeted cost of work performed

CPI_a = average cost performance index

SPI_a = average schedule performance index

This provides a more accurate model of the EAC by considering schedule performance as a factor that affects cost performance. It is especially useful for tempering a CPI in excess of 1.0 when the SPI is less than 1.0. This version considers that even if the task or project is getting the most for each dollar spent, but failing to achieve the schedule amount of work, it will impact the estimated value at completion (EAC).

It should be noted that an insufficient amount of established tracking data will reduce the accuracy of any forecasted EAC regardless of the method used.

Estimate to Complete

A second equally important calculation is for determining the amount of money (or labor-hours) required to complete the task (or project) if conditions remain as is. This is called the *Estimate to Complete* (ETC). The ETC is represented by this formula:

$$ETC = EAC - ACWP_c$$

where

EAC = the Estimate at Completion as derived from one of the methods shown earlier

$ACWP_c$ = cumulative actual cost of work performed

The ETC allows the project manager to determine what will be needed to complete the work from a monetary or labor-hours perspective. In short, what is the cash (or labor) requirement to complete the work? The team should keep in mind that this is a forecast and only as accurate as the values used to calculate it. If the EAC is weak or calculated from data that is too early in the life cycle of the task, the resulting ETC may be inaccurate.

Variance at Completion

Another simple calculation that is useful is whether the task (or project) will overrun or underrun the budget when it is complete. This calculation is called the *Variance at Completion* (VAC). The VAC is calculated by the following formula:

$$\text{VAC} = \text{BAC} - \text{EAC}$$

where

BAC = Budget at Completion from the baseline

EAC = Estimate at Completion derived from one of the methods previously shown

The VAC is a popular calculation among senior management that is always looking to add to the bottom line. A positive VAC tells the project team that the costs are under budget, while a negative value means the costs have exceeded the current target budget. Positive values, especially near the end of the project are often added to the estimated profit margin.

To-Complete Performance Index

The last factor used in forecasting future performance is an index that reflects the amount of value each remaining dollar in the budget must earn to stay within the BAC. It is called the *To-Complete Performance Index* (TCPI). It is based on a ratio of the remaining work to the remaining cost and is represented by this formula:

$$\text{TCPI} = \left[\left(\text{BAC} - \text{BCWP}_c\right) \div \left(\text{BAC} - \text{ACWP}_c\right)\right]$$

where

BAC = Budget at Completion from the baseline

BCWP_c = cumulative budgeted cost of work performed

ACWP_c = cumulative actual cost of work performed

The TCPI is the recommended performance from the status date going forward. For TCPIs greater than 1.00, there is more work remaining then there is budget to pay for it. TCPIs less than 1.00 indicate that there will be surplus funds in the budget after the work has been completed. Depending on the amount of project time left, TCPIs in excess of 1.20 may represent an unattainable level of performance before the project schedule expires.

TCPI is another big picture tool that is a favorite of senior management. Its accuracy rests with the accuracy of the cumulative data it is derived from. While it does not explain the problem it does forecast the outcome.

Having now explored the formulas for the EAC, ETC, VAC, and TCPI, the "four horsemen of the apocalypse," let's try an example of forecasting trends. The collected data for Task A is as follows:

$$BAC = \$150{,}000$$
$$BCWS_c = \$80{,}000$$
$$BCWP_c = \$82{,}000$$
$$ACWP_c = \$78{,}000$$
$$CV = \$4{,}000$$
$$SV = \$2{,}000$$
$$CPI_a = 1.05$$
$$SPI_a = 1.025$$

Here is how the given data and the previous formulas can be used:

$$\text{Simple EAC} = BAC \div CPI = \$150{,}000 \div 1.05 = \$142{,}857.00$$

or

$$EAC_{CPI} = ACWP_c + \left[\left(BAC - BCWP_c\right) \div CPI_a\right]$$

$$EAC_{CPI} = \$78{,}000 + \left[\left(\$150{,}000 - \$82{,}000\right) \div 1.05\right] = \$142{,}712.00$$

or

$$EAC_{com} = ACWP_c + \left[\left(BAC - BCWP_c\right) \div \left(CPI_a \times SPI_a\right)\right]$$

$$EAC_{com} = \$78{,}000 + \left[\left(\$150{,}000 - \$82{,}000\right) \div \left(1.05 \times 1.025\right)\right] = \$141{,}197.00$$

It is easy to see that all three numbers are reasonably close in value (approximately 1.1 percent), especially for a forecast.

From the aforementioned data, let's apply the formulas for the ETC, VAC, and TCPI:

$$ETC = EAC_{com} - ACWP_c = \$141{,}197 - \$78{,}000 = \$63{,}197$$

$$VAC = BAC - EAC_{com} = \$150{,}000 - \$141{,}197 = \$8{,}803$$

and

$$TCPI = \left[\left(BAC - BCWPc\right) \div \left(BAC - ACWPc\right)\right]$$

$$TCPI = \left[\left(\$150{,}000 - \$82{,}000\right) \div \left(\$150{,}000 - \$78{,}000\right)\right] = .94$$

Considering all of this data, the project team should be pleased with the all-around performance of this task.

It is essential to reiterate that all of the previously shown forecast calculations have a common theme, and that is, they are based on no changes to the current conditions and that past performance is representative of future performance. Any change of the status quo will impact the forecasted values.

It should also be noted that, despite the word "completion" found in EAC, ETC, VAC, and TCPI, they can all be calculated for various percentages complete of the task or project. The process is the same. So, for example, to calculate the expected values at 75 percent complete, the project manager would use the following:

$$ACWP_{75} = \text{cumulative cost of work performed at } 75\%$$

$$BCWP_{75} = \text{cumulative budgeted cost of work performed at } 75\%$$

$$BCWS_{75} = \text{cumulative budgeted cost of work scheduled at } 75\%$$

Project Percentage Analysis

Another class of performance measurement is for an overall status calculation. It is used for comparing the planned percent complete to the earned percent complete to the percent of cost. It is traditionally reserved for overall project analysis but also lends itself to task analysis. This factor can be calculated at any point in the process so long as the data is current with the status date. Data that lags the status date for any of the three input values will provide skewed results and misleading indicators. The formulas are as follows:

$$\text{Percent Schedule (Planned)} = (BCWS_c \div BAC) \times 100$$

$$\text{Percent Complete (Earned)} = (BCWP_c \div BAC) \times 100$$

$$\text{Percent Spent (Actual Cost)} = (ACWP_c \div BAC) \times 100$$

The formulas can be used in practice with the following given data for a specific status date:

$$BAC = \$150{,}000$$

$$BCWS_c = \$80{,}000$$

$$BCWP_c = \$82{,}000$$

$$ACWP_c = \$78{,}000$$

Therefore:

$$\text{Percent Schedule (Planned)} = (\$80{,}000 \div \$150{,}000) \times 100 = 53.33\%$$

$$\text{Percent Complete (Earned)} = (\$82{,}000 \div \$150{,}000) \times 100 = 54.66\%$$

$$\text{Percent Spent (Actual Cost)} = (\$78{,}000 \div \$150{,}000) \times 100 = 52.00\%$$

From this data the project manager can conclude that, as of the status date, the project was scheduled to be 53.33 percent complete, but actually determined by measurement to be 54.66 percent complete. Additionally, the team spent 52 percent to achieve 54.66 percent completion at that date. While gains were modest, they were positive. Overall, it is a favorable report.

Why All the Analysis?

By now you have to be wondering. . . why all the analysis? What is the end game? Well, it's simple. Construction companies need to make a profit. Most construction companies are financially fragile. That statement bears repeating for emphasis: *most construction companies are financially fragile*. Construction in general is a high risk business. One bad job can erase years of hard work or worse. It differs from manufacturing or sales in that each and every project is different. There is little uniformity. A contractor can build the same office building in multiple cities, on different sites, and at different seasons of the year and have an entirely different set of conditions to manage, and ultimately a different project. They receive payment *after* work has been done and the funds expended, they are subject to a high level of regulation, with constantly increasing prices for material and labor. The conditions under which they are expected to maintain quality control are less than good, let alone ideal! Weather is a factor. They contract with independent players that have their own agenda and managing them can be analogous to herding cats. Despite all of these inequities and challenges, each project must still generate a profit to invest back in the business.

Contractors have a limited number of opportunities to obtain work within their fiscal year. Compounded by the fact that they have limited resources to execute that work, all against a backdrop of the previously noted challenges. Each project therefore must contribute to the financial well-being of the company.

When a project is bid, a contractor includes an *estimated profit*, typically a percentage (it can also be a lump sum). From the moment the project is awarded, the contractor is looking for ways to ensure that the estimated profit is achieved. The best way to do that is to control both costs and time, because ultimately time is money. Projects that go long may prevent the contractor from securing other work, through lack of resources or even bonding capacity. Contractors try to increase the profit over and above what was estimated, as a means of providing a cushion or hedge against risk that may occur during the project.

Other profits streams come from negotiated buyout of subcontractors, material suppliers, and vendors. This is called *negotiated profit*, and it is a common practice. Over the life of the project, money will flow in to the negotiated profit column and it will flow out. It will be used to fill in omissions in the estimate, or for shortfalls in pricing due to increases. It is used to make up the difference between what was estimated and what it really cost. The goal is to have a net positive dollar amount in the negotiated profit column at completion.

Another source of profit is *management profit*. It is the money saved through proper and skilled management of the project. It starts with the crews performing work all the way up to the project manager. Seemingly insignificant savings can add up, such as:

- Efficient use of labor in self-performing work
- Avoidance of material waste, damage, or theft
- Planning to have appropriate quantities of material or the proper equipment when needed to avoid idle labor forces
- Small things like demobilizing the trailer and moving the office inside the building
- Returning rented equipment promptly when no longer needed
- Accelerating the work using methods that do not increase cost such as enlarging the crew of a task when there is sufficient work areas available
- Combining services such as clean up or scaffolding owned by each trade to one source
- Minimizing on site meeting times by using virtual platforms

The ultimate goal of management is to shorten the project schedule. Shaving two or three months off a 30-month schedule can reduce project and home office overhead significantly. In a fixed price contract, the savings go back to the contractor, and those savings add up.

All of the added profit streams do not happen by accident, they are part of the plan by management that requires the buy-in of all team members. It comes from being proactive, instead of reactive to day-to-day events.

In the end, all of the profit streams add to become the *targeted profit*. This will increase and decrease throughout the life of the project, but will become more stable near the end. Calculations such as ETC, EAC, VAC, etc. are designed to pinpoint what the net targeted profit will be!

Summary and Key Points

One of the tangible benefits of the comprehensive collection and analysis of data is its use in recognizing trends and forecasting future performance. There are several key values that aid the project team in predicting future cost and schedule performance. The main one is the EAC, which defines the anticipated cost based on current and past performance. The ETC is the amount required to complete the task, measured in dollars or in labor-hours. The VAC is the variance at completion, and it is the difference in dollars or labor-hours between the baseline value and forecasted value. These values can be calculated not only at completion but also as various percentages of completion of the project.

The To-Complete Performance Index (TCPI) is a ratio of work remaining to remaining cost. Another overall status frame of reference is project percentage analysis, which is the measure of planned earnings and costs.

All of the analysis and forecasting is to ensure that a contractor makes a profit on the project. Contractors have limited time and resources in a fiscal year, so assuming the risk of a project with no return for that risk is a fool's errand.

Chapter 10 Productivity will explore performance and some of the factors that affect it.

Key points of this chapter are:

- Repeated patterns of performance on a task are called trends.
- Accurate collection and analysis of data are essential to forecast future performance.
- Calculations of ETC, EAC, and VAC are all tools for forecasting how the project will finish financially.
- TCPI and Project Percent Analysis are big picture tools for high-level analysis and forecasting.
- All of the analysis is to estimate the profit the project will contribute to the company at the end of the project.
- The estimated profit can often be improved over the project life cycle.

Chapter 9 Recognizing Trends and Forecasting Performance: Questions for Review

1. Trend analysis can be used to forecast future labor requirements. True or False?
2. There are four considerations that trigger when data becomes an effective basis for forecasting. Define them.
3. What are trend charts used for? Give an example.

4. One of the most sought-after forecasted values is the total anticipated cost of the project when it is complete. This is called?
5. The amount of money required to complete the project under the current terms and trend is called?
6. The Budget at Completion (BAC) less the Estimate at Completion (EAC) is called?
7. A negative VAC tells the project team that the costs are under budget, while a positive value means the costs have exceeded the current target budget. True or False?
8. The TCPI is the recommended performance from the status date going forward. For TCPIs greater than 1.00, there is more work remaining than there is budget to pay for it. True or False?
9. Is an SPI of 1.023 acceptable? Explain.
10. What basic information can be derived from an SPI of 1.02 and a CPI of 0.97?

CHAPTER 10

PRODUCTIVITY

For most construction projects, the biggest concern or focus of project control analysts is labor. While the management of materials and equipment is crucial, labor has been and will remain the deciding factor in most projects. When we discuss efficiency ratios like CPI and SPI, we are almost always looking to productivity as the influencing factor. This chapter will discuss the concept of productivity and the factors that affect it.

Understanding Productivity

Productivity, as it relates to the construction process, can be loosely defined as the measure of the efficiency of the work being produced. It is the measure of output or return for each unit of input. To be more precise, it is the amount of work (output) produced per unit of labor (input). Productivity is the rate of work as measured per unit of time, or, in other words, units of work per labor-hour. Productivity in both private and government published data is expressed in this way. Whether it is noted as units/labor-hour or labor-hours/unit, it is still a measure of efficiency.

Productivity in construction is not uniform and does not remain constant over the execution of a task. Productivity in most applications is based on the average production of the crew or individual for the workday. Daily outputs for crews or individuals, used for estimating, are developed over time and are referred to as historical data. It is this data that represents the average productivity for the task under specific conditions. Most contractors track and analyze the productivity of the self-performed work that they do. This is mainly due to the high risk surrounding self-performed work. The actual productivity is compared to the estimated productivity and if the historical data needs to be adjusted, it is. This helps refine the estimated value for the next time the task is estimated.

In many manufacturing applications, productivity does remain constant because it is often automated. This is not the case in the construction industry. It is, however, the goal of management in both office and field to stabilize productivity to the point that it can be repeated with regularity and reliability.

Project Control: Integrating Cost and Schedule in Construction, Second Edition. Wayne J. Del Pico.

Frequently, the terms "production" and "productivity" are taken to be synonymous. This is incorrect. Production is the measure of output (units of work produced), whereas productivity is the rate of production (units of work per unit of time). Given this distinction, it is possible for a task to meet planned production for a workday while not achieving its planned productivity.

For example, consider a utility contractor that achieves the plan to install 100 feet of water line in a day, but requires twice the labor to do it. The contractor would be achieving 100 percent of the production, but only 50 percent of the expected productivity. Therefore, production and productivity are not reciprocal. Since contractors are paid by performance, not necessarily by direct hour of labor, productivity is related to earned value and ultimately the profitability of the contractor.

To accurately discuss productivity, a *production model* is a helpful visual tool. A production model is a mathematical expression of the production process that is based on collected data, measured in the form of quantities of inputs and outputs. In the construction industry, production is typically measured at the execution level or in the field. Here is an example of a production model for the water line installation that was mentioned earlier where LF is linear feet:

$$\text{Daily Output} = 100 \text{ LF} / 8\text{-hr workday}$$

To further expand the same model, consider the inputs that created the daily output to arrive at the productivity. If the crew consisted of two laborers, each working an 8-hour workday, then:

$$\text{Productivity} = 100 \text{ LF} \div 16 \text{ labor-hours}$$

or

$$\text{Productivity} = 6.25 \text{ LF per labor-hour}$$

So, for every labor-hour expended (input), the contractor's return is 6.25 linear feet of water line (output).

Productivity Index

As with all phases of project control, productivity is about the comparison of actual values to planned values. When the project was estimated, the labor portion of the estimate was based on a specific daily output by a specific crew under a specific set of conditions. It is the responsibility of the project team to analyze productivity as part of the control process. The comparison of planned output to actual output is called the *productivity index* (PI) and is represented by the following expression:

$$\text{Productivity Index (PI)} = \text{planned output} / \text{actual output}$$

Or, as an arithmetical formula using work hours:[1]

$$PI = BWH \div AWH$$

where

BWH = Budgeted Work Hours

AWH = Actual Work Hours

For example, if Task A had a cumulative total of 1,234 labor-hours at completion and the total budgeted labor-hours are 1,300, then PI is calculated using the following formula:

$$PI = 1{,}300\text{ L-hrs} \div 1{,}234\text{ L-hrs} = 1.053 \approx 1.05$$

A productivity index of 1.05 in the previous example indicates that the crew performed Task A 5 percent more efficiently than anticipated when Task A was estimated.

The productivity index is most accurate when used at the cost element level of the CBS, defined in Chapter 2 Introduction to Project Control. This is especially true when applied to labor cost and hours. At the cost account level or above, the PI is influenced by the material and equipment costs in the cost account summary. While this may be acceptable in some applications, it is not quite a pure snapshot of labor productivity. When the root cause analysis of a variance points to labor, it is recommended that the PI be calculated using, first, cumulative labor-hours, and second, cumulative labor dollars. The distinction is that labor *hours* are not impacted by wage increases whereas labor *dollars* are.

The PI can be calculated for any date in the scheduled duration of a task. The project manager need only determine the anticipated or budgeted number of labor-hours and the actual labor-hours accumulated to that date. The calculations of the PI in the early stages of completion (5–10 percent) may be less than accurate due to the learning curve process and should be reported with some tempering. The same is true for the final stages of the project (90–100) as production will be less than the efficiency of earlier stages due to wrap up, punchlist, etc. A review of Figure 6.1 in Chapter 6 Integrating the Schedule and the Budget may be helpful here. It is really the PI across the entire duration of the task that matters.

Despite the fact that productivity does not remain constant over the life of the task, estimators make the assumption it does, by applying an average output across the duration of the task. Aside from the task itself there are a wide variety of other factors that affect the productivity of a task.

[1]It should be noted that there is another common form of PI, which is AWH/BWH. Either form is acceptable so long as the practitioner understands the results.

Factors Affecting Construction Productivity

As mentioned on previous occasions in this book, the major deciding factor in the success or failure of almost any construction project is the successful management of labor. Labor is the quintessential "wild card" in the construction hand. The reason is its susceptibility to all sorts of factors with seemingly little, if any, warning or provocation. Minimizing fluctuations in productivity by managing the controllable aspects affecting labor is no small feat.

A bevy of previous research has attempted to identify and categorize the wide range of factors affecting construction productivity. In general, the factors can be separated into external (uncontrollable) and internal (controllable) factors.

Controllable Factors Affecting Productivity

Controllable factors are those that the project team can influence through proper management, planning, incentives, and coordination. They include:

- Ambient conditions within an enclosed space
- Proper planning and task analysis so the team understands what is expected of them
- Sufficient materials on hand to perform the work without interruption
- Avoidance of repeated interruptions or changes to task scope
- Minimizing of disruptions to work flow by crew changes
- Efficient relationship of staging area to work area
- Proper tools and equipment to perform the task
- Proper maintenance/repair/replacement of tools and equipment as required
- Effective training and familiarity with the task
- Positive leadership or supervisory skills on the part of managers
- Limits on absenteeism (including late starts and early departures)
- Clear communication of instructions to the workforce
- Encouragement of initiative on the part of workforce personnel
- Efficient setup and breakdown for work
- Clear understanding of scope and limits of work to perform
- Crew size and skill level appropriate for tasks
- Discouragement of substance or alcohol abuse on (or off) the job
- Elimination of unsafe working conditions
- Consideration of coworker compatibility issues
- Mindfulness of technique; means and methods

Uncontrollable Factors Affecting Productivity

There are many factors affecting productivity that, despite all of the professional management and planning skills, are beyond the manager's control. Some of the more well-known ones include:

- Personal ambivalence or apathy of the worker
- Outside weather conditions: precipitation, temperature, etc.
- Personal problems (unrelated to work) diverting attention from work
- Minor health problems: colds, fatigue, aches and pains
- Economic conditions in the region
- Strikes or work stoppages
- Material shortages or unavailability
- Frequent scope changes by owner/client

One can clearly see that the controllable factors far outweigh the uncontrollable ones.

Recommendations for Improving Productivity

Not every moment of the workday is spent producing tangible work. According to multiple industry sources, a typical 8-hour shift for a journeyman is only 61–66 percent productive (the average percentage of time spent doing the primary work). The remaining time consists of the following:

- Studying plans (3 percent)
- Material procurement (3 percent)
- Receiving and storing of materials (3 percent)
- Mobilization (5 percent)
- Site movement (5 percent)
- Layout and marking (8 percent)
- Cleanup (3 percent)
- Breaks and nonproductive time (6 percent)

This means up to 39 percent, over one-third, of the shift is spent doing things other than the primary task. However, many of these things are an essential part of the work process. The goal of the successful management of labor is to reduce or remove impediments to improve the 61–66 percent. The following are some general recommendations for improving productivity:

- Plan appropriately; schedule work whenever possible to take advantage of the weather or any ambient condition.
- Order materials and equipment and confirm delivery before scheduling workforce personnel.

- Ensure that supervisors, foremen, and tradespersons understand the full scope of the task and what is expected of them.
- Implement workforce management procedures and measure communication and performance.
- Minimize interference and disruption to production cycles.
- Encourage initiative and cooperation among workforce personnel.
- Maintain safety equipment and promote a safe work environment.
- Whenever possible, ensure that staging areas are as close to work areas as practical.
- Ensure that personnel are adequately trained for the task.
- Provide proper tools and equipment to perform the task.
- Provide sufficient crew personnel and mix for the task.
- Promote pride and ownership in workmanship.
- Promote quality assurance practices and procedures to reduce re-work.

Occasionally, there is no practical way in which to improve productivity. There are a number of reasons for this. The estimate could be too optimistic. Management could be incorrect in their expectations of productivity for the task being performed. These are among the most common.

Another is the poisoning of a crew dynamic by a supervisor or an individual tradesperson. This too is more common than one would think. Despite assurances that the crew is trying their best, there is a covert rebuke of any attempt to improve productivity.

Either condition can be sorted out by measurement and analysis. This may require that the management team conduct a time study of the task. A time study consists of assigning an individual to observe and track the production for a definitive period of time. Production during the time period under observation can be compared with earlier production rates when not observed. This often reveals the cause of the anomaly.

Lastly, no discussion of improving productivity would be complete without addressing the topic of incentivizing the workforce. There are numerous ways to incentivize a construction workforce, but nothing works like money. . . specifically a bonus. Bonuses are often offered and paid for exemplary performance when management feels it is warranted, to complete a phase or meet a deadline. The problem with bonuses is that they come to be expected. When they are not offered, it creates a feeling of being undervalued amongst the crew. This can create resentment leading to a disgruntled crew and reduced performance. It is human nature. The result may be that good employees are no longer dependable to do their days work. This leaves no option but to replace them.

The use of bonuses should be done sparingly, if at all.

The Use of Overtime and Premium Wages

The use of overtime and the payment of premium wages is frequently turned to as a way to improve production. It can often bring a project back on schedule or increase production of a task, but at what cost? What considerations are appropriate in the decision to turn to overtime? Chapter 11 Acceleration and Schedule Compression will consider this topic in more depth as it is a means of acceleration.

Summary and Key Points

Productivity is a measure of the efficiency of labor. The productivity index compares planned production to actual production. There are numerous factors that affect productivity, many of which are within the control of the project team and most of which center around management and conditions of the worksite.

Some factors are outside the control of the management team and should be addressed with the specific worker or crew. The reason for reduced or abnormal productivity for a task can be analyzed and determined via a time study.

Productivity indexes measure the efficiency of the labor and should be analyzed at the cost element level so they are not influenced by material and equipment costs.

In Chapter 11 Acceleration and Schedule Compression, the discussion will focus on the decision-making process of trading costs to accelerate performance.

Key points of this chapter are:

- Productivity is a measure of labor efficiency.
- It is possible to meet the production without meeting the productivity.
- There are many factors that affect productivity some within management's control and some outside.
- Management should strive to improve productivity by addressing the controllable factors.
- The efficiency of a task is not consistent over the life of the task but it is estimated as a uniform productivity.
- Financial incentivizing can often create problems on future projects when the workforce feels it should be incentivized and management does not.

Chapter 10 Productivity: Questions for Review

1. Explain the difference between productivity and production.
2. Productivity is the measure of output or return for each unit of input. True or False?
3. Productivity in construction is uniform and remains constant over the execution of a task. True or False?
4. What is a production model and how is it used?

5. The comparison of planned output to actual output is called?
6. Explain what a productivity index (PI) of 1.034 indicates.
7. Factors that affect productivity can be separated into external (uncontrollable) and internal (controllable) factors. True or False?
8. Statistics show that crews are productive about 95 percent of the workday. True or False?
9. Why are labor-hours a better basis for a productivity index (PI) than dollars?
10. Identify five controllable and five uncontrollable factors that can affect productivity.

CHAPTER 11

ACCELERATION AND SCHEDULE COMPRESSION

Construction projects always have delivery dates. Some are dictated by the owner, and some are dictated by the contractor. Most delivery dates are contractual. There are a multitude of reasons, but the most prominent one is that a project needs a delivery date for control. Without a delivery date, there is no goal and human nature's proclivity toward procrastination takes over. In construction, the age-old adage of "time is money" becomes very apparent for all parties when the work is delayed. Many contracts include the phrase "time is of the essence" to establish the criticality of time management as it relates to economic significance. Lost revenue, extended overhead, additional financing, and pending damages can all change the dynamics of the project and, as such, must be managed very carefully. Beyond money, there are the logistical considerations that are part of a delay. What happens to a restaurant grand opening that is rescheduled due to construction delay or the first day of school that has to be postponed? Both have a far-reaching impact.

Schedule is important and, on occasion, more important than cost. Every project manager will at one time or another face a project that has fallen behind schedule. The project manager may have to consider acceleration of the work to make up the time or may be requested to compress the schedule to deliver early. Both are fairly common and must be understood fully to manage and control.

Acceleration

Delivering on time or ahead of schedule is important for all parties, from subcontractors all the way up to the owner. Finishing within or under budget is another important goal. The real achievement, however, is doing both simultaneously. Unfortunately, this may not always be possible; in fact, it is fundamental knowledge that shortening the schedule costs money, sometimes significant amounts of money. At the point where an accelerated pace of the work is diagnosed as the cure for delay, the project manager may have to make the decision to spend money.

Project Control: Integrating Cost and Schedule in Construction, Second Edition. Wayne J. Del Pico.

Acceleration is defined as the addition of resources or work hours to a task or project. Acceleration results in *schedule compression* or reducing the duration of the task or shortening the entire project without diminishing the scope of the project. For the project schedule to be compressed, the acceleration must be applied to the critical path or a critical task. To accelerate a noncritical task rarely benefits the schedule. However, it should be remembered that if ignored long enough, all tasks can become critical and end up on the critical path.

While a small amount of delay can be expected on all but the simplest, most repetitive types of projects, the delay of too many tasks can end in a complete abandonment of the project schedule. It results in trades working on top of each other, damage, or the compromise of quality, irresponsible expenditures, and general anarchy and chaos. It always ends in delays to milestones or, worse, delays to Substantial Completion. In the final tally, it produces mistakes that inevitably have to be corrected. Avoiding excessive delay is one of the principal purposes for project control.

There are three different types of acceleration that the average project may potentially encounter. Acceleration is issued in the form of written direction from the owner to the contractor, or from the contractor to the subcontractor, and can be classified in three categories: actual, constructive, and forced.

Actual Acceleration

The first of the three categories is *actual acceleration*. Actual acceleration is the directive to complete the project ahead of the accepted or contractual schedule. This directive comes from the owner to the contractor through the architect, and is then passed on to the subcontractors. It also implies that compensable acceleration costs will be borne by the owner via a change order. When mutually agreed upon, this is fairly straightforward. Actual acceleration is based on the premise that the critical path is correct or, more accurately, reasonable, and to compress the schedule any further will require additional work hours either in the form of daily overtime, a second shift, or weekend work.

Actual acceleration requires an in-depth analysis of the remaining work, remaining duration, and the available resources. It is no small challenge and often requires the input of other team members, especially the subcontractors that will perform on the accelerated schedule.

Actual acceleration may go beyond just labor, it may include costs for expediting materials. This can include accelerated offsite fabrication costs, dedicated shipping costs, or on occasion an airplane ride. This is not just for the GC, but also for subcontractors. Since changing the Substantial Completion date is a contract modification, it requires a change order. Change orders are final once executed, and do not allow for a "second bite of the apple" if the acceleration is not calculated correctly. It should be noted that the accelerated date is not a "do the best you can" date. It is a firm contractual requirement and if the contractor and subcontractors fail to achieve this date, it may result in the owner being relieved of paying for the change order. When GCs execute change orders for actual acceleration with their subcontractors, it should mirror the language of change order between the GC and the owner so that the flow down provisions of the change order remain intact.

Constructive Acceleration

The second category of acceleration is called *constructive acceleration*. Constructive acceleration occurs when the contractor requests an extension of time and is denied. If the delay that resulted in the request for an extension of time is an excusable delay, the extension should be granted. A classic example results from the issuing of change order(s) near the Substantial Completion date, without allowing an extension of time for the contractor to perform the work.

Unfortunately, it may not be possible to grant the time extension, and the contractor may be required to accelerate progress so as to achieve the original date. This will typically result in approved change orders being performed at additional costs due to premium wages. Constructive acceleration exists when the following conditions occur:

- There is an excusable delay to the project.
- The contractor requests a justifiable extension of time for the excusable delay.
- The request for an extension is denied.
- The contractor is directed to achieve the original completion date.
- The contractor actually accelerates the work to meet the original date.
- The contractor incurs additional costs for acceleration.

If these conditions exist, the contractor may have a valid claim for additional costs as a result of constructive acceleration.

Often owners release multiple change orders at the same time, near the end of the project (95–99 percent complete). This can be because they are unsure of remaining funds to pay for the changes and must wait till the end of the project or in an attempt to be punitive to the contractor. In this case it is done to trip up the contractor with the hopes they will miss the Substantial Completion date and trigger damages. The project manager should keep in mind that change proposals are submitted at specific times in the schedule. The fact that an owner may sit on them for weeks or even months may impact the cost of the change order when they are finally authorized. If the conditions under which the change proposal were offered have changed to the point that the change proposal is no longer valid in its current state, the contractor should not accept it. Again, four change orders issued in the last two weeks of the project when "all hands are on deck" already, will exacerbate the resource over-allocation. If these change orders are not on the critical path, and consequently not required for Substantial Completion, then they should be postponed until *after* Substantial Completion has been achieved. The project manager must document their actions with the owner and respective subcontractors thoroughly in this case.

Forced Acceleration

There is also a need for acceleration when delay occurs as a result of nonperformance. This type of acceleration is sometimes referred to as *forced acceleration*. This occurs frequently when a subcontractor on the critical path is falling behind. While this is not acceleration in the truest sense of the definition, it does require an increase in production to recover the schedule. This type of acceleration is typically not compensable. It also does not always have the cooperation of the underperforming subcontractor. Most project managers would prefer to never pay acceleration costs to an underperforming subcontractor; however, one must acknowledge that it should be a business decision and not an emotional one. To spend $5,000 today to avoid $20,000 in liquidated damages at the end of the project makes sense to even the novice. This is especially true if the $20,000 in liquidated damages cannot be deducted from the balance owed to the subcontractor responsible for the delay. If the subcontractor is unable to accelerate the work for financial reasons, but can guarantee schedule recovery with the acceleration, it may be worth considering.

Formal acknowledgement of being behind schedule by the defaulting party is rare due to the downstream consequences. Admitting that they are the cause of the delay can open a flood gate of damages in the form of acceleration costs for the subcontractors that follow. Acknowledged or not, the project manager must act to correct the delay. The cooperative subcontractor or nonperforming party should be allowed to concur with and assist in the development of the acceleration plan. The project manager should also remember that most contracts provide for the augmentation or supplementing of a subcontractor's workforce if the subcontractor is failing to remedy the delay by accelerating the work. Caution is urged when supplementing a subcontractor's workforce as it is viewed as an invasion of the subcontractor's proprietary business and may exacerbate an already dire situation, including abandonment. Regardless, it is an option that must be considered. Care should be taken when augmenting a subcontractor that has a separate license and permit as it can result in them removing their license from the permit as they abandon the project. This can end in a project with no plumbing or electrical permit.

It should be noted that each of the forms of acceleration is not limited to the relationship between the owner and the contractor. Acceleration directives can, and frequently do, occur between the contractor and the subcontractor. It should be further noted that occasionally the contractor decides to deliver early without the direction or compensation of the owner. In this scenario, the contractor will pay acceleration costs to the subcontractors and vendors as well as premium wages to employees to deliver early.

Most seasoned project managers know that schedule slippage does occur on every project to some degree. Many proactively consider phases or tasks that can be accelerated even in the early stages of the schedule or in the planning phase. This provides a preconceived plan for recovering the schedule to be ready when needed. It is also important to follow any and all contractual requirements for notification to the delaying party. Without proper and timely notification in accordance with the contract, the contractor may be on shaky ground for the recovery of acceleration or augmentation costs.

Lastly, the project manager is advised to secure in writing from the architect, with the owner's concurrence, a directive to accelerate the work that stipulates exactly what costs are compensable. Most acceleration techniques are labor-based, and, as a result, costs can go exponential quickly.

Schedule Compression

The decision to accelerate work and incur costs must have a tangible benefit to the project. Schedule compression is not always about the shortest duration of the task. It is more about the balance of getting the shortest duration for the least cost. In colloquial terms, it is about the "biggest bang for the buck." Schedule compression is commonly defined as modifying the plan and/or schedule to complete critical tasks or the project early. Schedule compression is the resequencing or rescheduling of tasks with minimal or no cost impact. *Crashing* the schedule, in contrast to compressing the schedule, is the addition of resources or accelerating the work to shorten the duration of the task or the project itself. Crashing the schedule is often viewed as a more extreme measure arriving at maximum reduction of the schedule. It can sometimes evolve into complete disregard of financial controls in favor of time and always has cost implications.

Since schedule compression is common, the techniques employed are part of the skill set of the professional project manager. The Construction Industry Institute conducted a study of schedule compression through Colorado State University and issued their findings in a report. This report, entitled *Concepts and Methods of Schedule Compression*, listed nearly 100 techniques for consideration when compressing a schedule. The techniques run the full spectrum from management functions to constructability to productivity improvement. The following are a sampling of some techniques with a history for shortening construction phase schedules:

- Timely review and response to submittals and documents
- Detailed review and analysis of baseline schedule for possible logic or duration flaws

- Improved materials procurement and delivery management including the identification of long lead items
- Additional production personnel added to existing crew
- Additional hours of work: overtime and weekend work
- Additional shifts: providing labor resources to work a second shift
- Special crews to manage and stock materials to work locations
- Application of new technology or techniques
- Provision of incentives for improved performance
- Location of project managers at the jobsite
- Improved and accelerated change order management
- Off-site fabrication to reduce installation time
- Maintenance of high workforce morale
- Constructability review in advance of performing work
- Scrutiny of value engineering proposals
- Substitutions based on availability
- Resequencing of tasks to take advantage of float
- Accelerating of tasks with the least acceleration cost

Time-Cost Trade-Off Analysis

The reasons for acceleration can be numerous and unique to each project; however, professional contractors do not as a rule enjoy spending money on acceleration without a distinct, tangible benefit. To that end, most project managers compress the schedule in increments or steps starting with the least expensive methods. To decide which tasks offer the most time-savings for the least costs requires careful analysis. This process is called *time-cost trade-off analysis*.

The analysis begins with a review of tasks on the critical path. The two criteria that often provide the best opportunity for the optimal time-cost trade-off are tasks with the longest duration and tasks that are labor-intensive. Short duration tasks are often difficult to compress without significant cost, if even possible. Schedule compression is a trial and error process, with many options being reviewed before one is selected and implemented.

Costs for schedule compression are a combination of the contractor's (and subcontractor's) direct and indirect expenses or costs. Direct costs are for production or overhead that is *directly related* to the individual project. For example, on-site material, labor, equipment, and subcontractors involved in the work are considered direct costs. So are the superintendent's salary and other project overhead costs because they are directly related to that one project. Indirect costs are for items that support or are ancillary to the production work, and are *indirectly related* to the project. Indirect costs are part of the contractor's cost to operate the company as a whole, rather than one specific project. Indirect costs are shared among all projects in proportion to their duration and cost. An example of an indirect cost is the rent for the home office.

All costs, direct and indirect, can be either fixed or time-sensitive. An example of a fixed direct cost is a building permit. The fee for the permit is paid once regardless of the length of time it takes to complete the work. An example of a time-sensitive cost is the superintendent's salary: the longer the superintendent is onsite, the more cost is incurred by the project. Other time-sensitive costs may originate in the home office. Home office overhead is indirect and almost always time-sensitive. As the schedule is shortened, it has the by-product of saving time-sensitive costs by default.

Time-Cost Trade-Off Concepts

As the schedule is shortened, the fixed and time-sensitive costs, both direct and indirect, are affected differently. As time decreases, so do time-sensitive costs. Any shortening of the critical path not only has the value of an improved schedule, but it has the benefit of reducing time-sensitive project overhead costs and potentially home office overhead as well. Direct costs for materials, labor, and equipment that are not time-sensitive, but are fixed due to quantity and available resources, are impacted as more work must be completed in a shorter window of time. Direct costs often increase with schedule compression through overtime expenses on labor or expediting of materials.

Time-cost decision making centers around two key factors:

1. The difference between the original duration for the task as identified by the baseline schedule called the *normal duration* and the accelerated duration referred to as the *crash duration*.
2. The difference between the original cost for the task as identified by the baseline budget called the *normal cost* and the accelerated cost referred to as the *crash cost*.

The ratio of cost incurred per day saved or the time-cost ratio can be expressed in the following formula:

$$\text{Time-Cost Ratio} = (\text{Cost}_c - \text{Cost}_n) \div (\text{Duration}_n - \text{Duration}_c)$$

where

Cost_c = crash cost expressed in dollars

Cost_n = normal cost expressed in dollars

Duration_n = normal duration expressed in workdays

Duration_c = crash duration expressed in workdays

For example, if the following data was provided for Task A, then

$Cost_c$ = \$7,600.00
$Cost_n$ = \$5,500.00
$Duration_n$ = 10 workdays
$Duration_c$ = 7 workdays

$$\text{Time-Cost Ratio for Task A} = (\$7{,}600 - \$5{,}500) \div (10 \text{ workdays} - 7 \text{ workdays})$$

$$TCR_A = \$2{,}100 \div 3 \text{ workdays} = \$700 \text{ per workday}$$

Once the candidate tasks have been analyzed, the tasks with the lowest cost per unit of time (workday) should be shortened first. As additional time is required, the project manager should select tasks with an increasing TCR. The only exception to this is when a crashed task with a lower cost per day has more risk (to be discussed in Chapter 13 Risk Management). Using this method of analysis, the project manager can work through the critical path tasks starting with the longest duration tasks until sufficient schedule compression has been achieved.

Indirect costs are linear in exposure. Most indirect costs are time-sensitive. As the work is accelerated, the duration decreases, and the indirect costs are reduced. This can be seen in Figure 11.1, which is a graph of indirect costs as a function of time.

Direct costs increase in a nonlinear fashion as the work is accelerated. The concept is that the more the work is accelerated, the more expensive is the cost of each saved day. This is partly due to the use of more expensive technologies or techniques to save time. Also, as noted earlier, the time-cost trade-off starts with the tasks that have the least expensive TCR followed by the

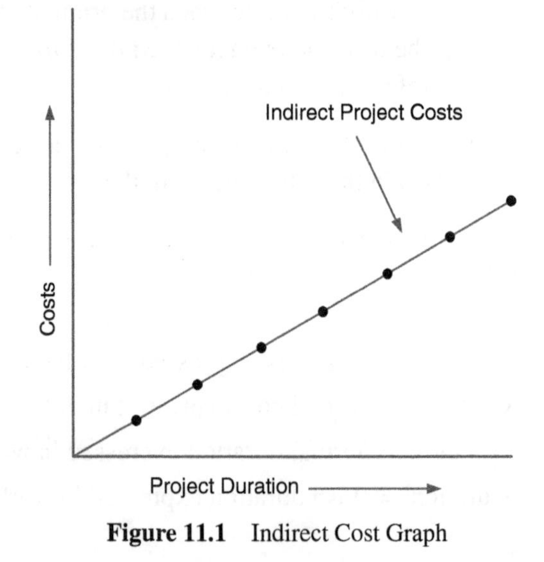

Figure 11.1 Indirect Cost Graph

selection of more expensive ones if more days are necessary. The costs are not linear; in fact, they can increase quite rapidly as the duration is reduced. Figure 11.2 is a graph of direct costs plotted as a function of time. Note that as the work is accelerated and duration decreases, direct costs increase.

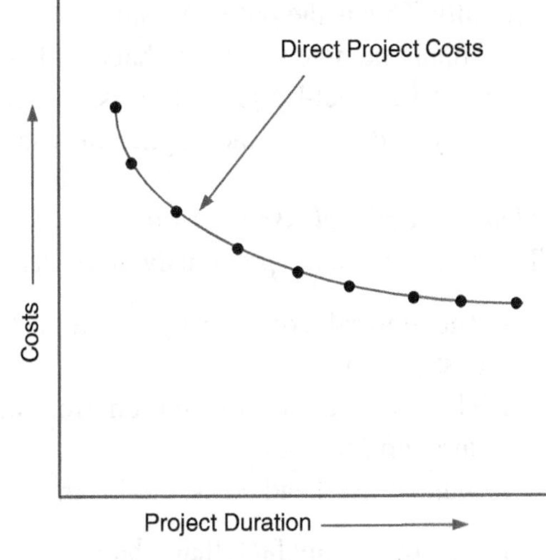

Figure 11.2 Direct Cost Graph

Direct and indirect costs can be combined to represent total project costs and then be plotted as a function of time. Figure 11.3 illustrates the total project cost versus project duration. The total project costs can be calculated for a series of different task (or project) durations to determine the *optimal point* on the graph. The optimal point is the lowest point in the curve and represents the *least cost* with the *least cost duration*.

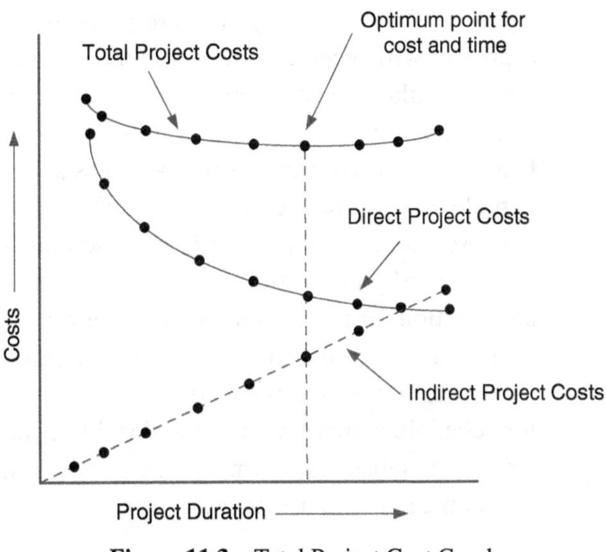

Figure 11.3 Total Project Cost Graph

The total project cost graph starts with normal duration and cost. As the work accelerates, the total costs decrease at a nonlinear (decreasing) rate, because the direct cost is increasing at a nonlinear (increasing) rate. Indirect costs are decreasing at a linear (constant) rate. Eventually the total cost curve will start to increase again due to direct costs increasing exponentially. This is the optimal point.

It should be noted that the shape and slope of the direct cost curve will differ on a project-by-project basis. Figures 11.2 and 11.3 are for illustration only, and not an exact replica of what can be expected.

Time-Cost Model Assumptions

The time-cost model previously discussed relies on three key assumptions:

- The normal cost ($cost_n$) for a task is less than the compressed cost ($cost_c$).
- The normal duration ($duration_n$) for a task is longer than the compressed duration ($duration_c$).
- Indirect overhead costs are linear.

It is a noteworthy fact, that when critical tasks are shortened, it can have the unintended result of changing the primary critical path by making non-critical or near-critical tasks critical. Before any schedule compression is implemented, the project manager is urged to review the changes and their impact on the primary critical path to ensure it is the same or that the newly revealed critical path is acceptable.

Making the Case for Acceleration

For many companies, the decision to accelerate the work and incur the associated costs is a decision that requires concurrence of executive management. Project managers are required to demonstrate the benefits that a project will receive from acceleration in clear, concise, and financial terms, similar to a cost-benefit analysis that one might prepare when considering a value engineering proposal. Too often the cost far exceeds the benefits returned or never fully becomes known until the task or project is complete. The practice of analyzing the financial impact and the benefits from schedule compression is an essential part of a project manager's duties, worthy of the time invested. This is called "making the case" for acceleration, or more formally, presenting a business case for the action. It should be noted that we are not referring to the occasional overtime needed to get a task finished or caught up. Presenting a business case for schedule compression is for sizeable commitments of money and resources to substantially impact the project schedule or reverse a negative trend on a major phase or task.

Experience has shown that developing a business case for acceleration and the ensuing discussion among team members or upper management can often pose questions or concerns beyond a single viewpoint. A plan that survives the challenges of senior management can often garner the support needed to ensure its success. A project manager who can calculate and present the case for spending $20,000 now to save $100,000 later is a valuable asset to any team or company.

One final note on acceleration: While it is almost always perceived as positive to finish early, the project manager is directed to discuss any schedule compression that would deliver the project early with the owner and the architect. Accelerating the work, even if the contractor bears the costs, will reflect in increased earned value per billing period in the Applications for Payment. Unprepared owners may be unable or unwilling to fund the increased cash being paid out, especially if they have structured a loan based on the original cash flow illustrated by the baseline.

In addition, an unprepared owner may be unable to accept the project due to budgetary or staffing concerns. Early delivery can also create concerns for warranties if their terms will partially expire before the project has been occupied. Early delivery before budgets are funded could leave a project unheated or without electricity, thereby voiding warranties for products with temperature and humidity thresholds.

Overtime and Premium Wages

It is a fairly common practice to turn to overtime work and the payment of premium wages when productivity wanes and the schedule falls behind. Understanding the implications of this technique beyond just its cost should be acknowledged before jumping in with both feet.

There are specific federal and state laws for the payment of wages when work exceeds 40 hrs in a week. For union trades, the threshold for overtime is 8 hrs. In either case, there is an added cost, in the form of *premium wages,* that the employer is required to pay the employee when those two thresholds are exceeded. The most common overtime wage scales are *time-and-a-half* and *double-time*. Time-and-a-half is one and one half the base wage and double time is twice the base wage for an hour's labor. In addition to wage, there are modifiers in the form of taxes and insurances that the employer is responsible to pay for the premium portion of the wage. All of these costs are easily calculated. The less straightforward calculation is the output that will be gained from the added wage and time. Remember that the goal is to achieve predictable outcomes from the added wage.

There are two basic classifications of overtime: *occasional overtime* and *scheduled overtime*. The term occasional overtime can be defined simply as work hours in excess of the employee's normal work week (or day). For example: the employee is asked to work an extra 2 hrs in a day to complete a phase or task. The second classification called scheduled overtime can be defined as the normal or regular workweek for the employee is routinely in excess of 40 hrs per week or 8 hrs per day every week. For example: the employee works five 10-hr days and 4 hrs on Saturday as their normal work week. The productivity outputs of both classifications can be quite different.

Numerous studies have been conducted and the research analyzed[1] and all conclude that occasional overtime in the construction industry has little if any impact on productivity. However, the same studies conclude that scheduled overtime, especially prolonged periods of scheduled overtime, have a dramatic effect on productivity. Reducing productivity to as little as 55 percent the normal productivity for the same hour. This is despite paying time-and-a-half or double-time for that hour. When productivity is averaged over a 4-week period the average productivity is less than 70 percent per week. Clearly, not a worthy investment. What other options does the project manager have?

Another method of acceleration is adding a second shift. This requires a second group of employees and more management for supervision. For many smaller companies with limited resources, this may not be an option. However, for those companies with the resources, a second or even third shift can get the work done with a *pay differential*. A pay differential is an increase in pay for unusual or challenging work conditions. It is often a percentage of the original wage scale added to the base rate. While this labor-hour costs more, it does not impact the productivity quite as negatively. Whatever the choice, the project manager and team should not embark on the use of premium wages without a thorough understanding of the consequences and the support of senior management of the company.

Recovery Schedule

In reality, project schedules slip. Part of the project manager's duties, once the slipped tasks and the responsible party have been identified, is to revise the schedule with the corrective measures. This type of revision to the schedule is called a *recovery schedule* and is a necessity for the successful completion of any project.

[1] A sample of studies supporting this conclusion: "Schedule Overtime Effects on Construction Projects" (Business Roundtable (BRT), 1980); "Overtime and Productivity in Electrical Construction" (the National Electrical Contractors Association (NECA), 1989); "Schedule Overtime and Labor Productivity: Quantitative Analysis," by Dr. H. Randolph Thomas, et al. (Penn State University, 1997); and "Modification Impact Analysis Guide," Publication No. EP 4150-1-3 (U.S. Army Corps of Engineers (COE), 1979).

Schedule recovery planning requires both the acknowledgement of the parties who are responsible and their cooperation in getting it back on track. The acknowledgement is more problematic than it might first appear. Any delay can have a chain reaction that affects all work and subcontractors downstream from the delay. Most, if not all, subcontractors are reluctant to own up to a delay issue for fear of the potential for financial repercussions. On the other hand, it is not within the project manager's authority to indemnify the nonperforming party from claims that may result. The most effective method of schedule recovery is to focus on a solution rather than finger pointing as to the cause.

Many contracts require that a contractor whose project has slipped the schedule by 10 percent or more, or has repeatedly failed to achieve schedule deadlines, must demonstrate through the preparation of a recovery schedule how they will get the project back on schedule.

It is the author's experience that many contractors are unwilling to label a plan a recovery schedule for fear of the legal consequences of acknowledgment that the project is behind. Regardless of the title, some modification or adjustment to the schedule is necessary to get the project back on track. The recovery schedule should employ one of the aforementioned accepted methods for compression.

Scheduling, especially CPM scheduling, is done almost exclusively by computer with sophisticated software. While the initial learning curve may be cumbersome, the time saved and the ability to try multiple scenarios for compression are enormous. (Remember before making any adjustments to always save a copy of the baseline, or current schedule.) However, when calculating the least cost duration, it should be done manually. That requires that for each day saved, the net increase in direct cost is subtracted from the decrease in indirect cost to arrive at a total cost per day. These calculations can often be performed with Microsoft Excel™ or similar spreadsheet applications.

Summary and Key Points

The requirement for schedule compression can occur for many reasons. To reduce the duration requires that the work be accelerated. There are three forms of acceleration: actual, constructive, and forced. Each has unique conditions that are required for its use. Deciding on which activities to crash is part of time-cost trade-off analysis in which the lowest cost per day task is crashed first, then proceeding to more expensive tasks. The exception to this is when lower-cost tasks offer more risk for failure. In this circumstance, higher-cost tasks are selected.

Making the case for acceleration requires careful and thorough calculations of savings and increased costs as well as any potential areas of high risk. There must be a distinct reduction in schedule for any funds expended. Change order are priced and issued independently. However, near the end of a project an owner may release a group of change orders that can change the schedule and even the cost from what was submitted. This should be explored before accepting the change orders.

The use of overtime work may affect more than the cost of the task itself. Prolonged periods of overtime reduce productivity and can fatigue a workforce.

Early delivery of a project is almost always desirable; however, discussion with and approval of other parties, especially the owner, are essential to avoid an unprepared or an unwilling owner.

Recovery schedules are often contractually required when a project exceeds established thresholds of delay. It should be noted that change orders to accelerate the work may be rejected if the goal is not achieved. In other words, there is no payment for partial performance.

Chapter 12 Resource Management will explore the timely and efficient use of labor and equipment resources as a means to control costs and schedule.

Key points of this chapter are:

- Projects often fall behind at various points in the schedule. Recognizing and managing the recovery through a modified schedule requiring acceleration offer the most effective method.
- There are three types of acceleration: actual, constructive, and forced. Each type is used differently and should be selected as the situation warrants.
- Deciding which tasks to accelerate may require a time-cost trade-off analysis resulting in the identification of which tasks cost the least for the most days gained.
- Overtime with premium wages should be used sparingly and for short periods of time only.
- Deciding on acceleration may require the management team to provide evidence of effectiveness for the money spent.
- Recovery schedules can be challenging to construct and execute and will require the concurrence of the participants to be successful.

Chapter 11 Acceleration and Schedule Compression: Questions for Review

1. Define the three types of acceleration.
2. What is an excusable delay? Provide an example and explain.
3. Explain why most contracts provide that a general contractor can supplement a subcontractor's workforce even if the sub is an independent entity.
4. Define the term "Schedule Compression."
5. Schedule compression rarely results in the increase of labor resources or labor-hours. True or False?
6. What is time-cost trade-off analysis?
7. Explain the difference between a direct cost and an indirect cost on a project.
8. Explain the term "crash duration."
9. It is the purpose of the time-cost model to identify the task with the least cost per crash day. True or False?
10. Why are contractors reluctant to produce a recovery schedule?

CHAPTER 12

RESOURCE MANAGEMENT

From the very beginning, this text has stressed the value of planning. Planning is not limited to the earliest phases of the project, but is the continual mantra throughout the entire process. Right up to the very end, the project manager must plan for the project closeout. Initial planning for baseline controls is considered the long-term planning aspect of the project. However, there is still a need to plan for resources to execute the work in the short term. This part of project control is called *resource management*. Resource management has two key goals: to ensure that the correct resource is available at the correct time and in sufficient quantity, and to ensure that the resource is used efficiently.

Shortages and mismanagement of resources are frequent causes of substantial delays. Most project managers have watched tradesmen standing idly by due to lack of materials, or experienced delays due to the wrong piece of equipment used in performing a task. Proper management of resources—materials, labor, and equipment—can minimize or even eliminate these types of inefficiencies.

Resources

In the construction industry, resources are defined as the labor, materials, and equipment necessary to perform the work of the contract. The three categories can be further explained.

Labor can be subdivided into two types:

- Salaried personnel typically consist of supervisory or management staff. They are paid a fixed salary exclusive of the commitment of time. They are not generally tied to a single task.
- Hourly wage personnel are tradespersons who are hired to perform a specific task based on their experience and skill set. They are paid for hours worked.

Labor is the primary resource and is of the greatest concern due to its flexibility and the fact that its performance is affected by a wide variety of

Project Control: Integrating Cost and Schedule in Construction, Second Edition. Wayne J. Del Pico.

factors. Labor is and always has been the predominant concern of the project manager when it comes to cost.

Material covers a wide range of items and can be classified as one of two types:

- Support materials are not incorporated in the work, but provide support for the work. An example is the formwork, tarps, platforms, and enclosure materials needed for weather protection.
- Installed materials are permanently incorporated in the work. Examples are concrete, bricks, lumber, and wire. This can be broadened to include equipment such as water heaters, roof top units, and permanent generators. For budgeting and estimating purposes, these types of equipment are classified as materials.

The third category of resources is the types of equipment used to perform work that have no permanent status in the project:

- Examples are cranes, excavation and backfill equipment, temporary power generators, and temporary heaters. Once the need has been satisfied, the equipment is removed from the project. It frequently has a direct correlation to labor, since without operators, much of the equipment remains idle.

Another key resource that has intentionally been omitted are subcontractors. Subcontractors play a key role in the performance of the work, however, the prime or general contractor has less control over this workforce due to the sub's independent contractor status. This will be discussed later in the chapter.

Resource Allocation

Scheduling and coordinating resources so that they arrive on-site as needed and are removed when not needed is called *resource allocation* or *resource loading*. When performed properly, it is the effective and efficient utilization of all resources. Shortages of material or scheduling the wrong piece of equipment will inevitably lead to a delay in performing the task and may delay the project itself. Allocating too many tradespersons to the crew will leave some underutilized and increase the cost per unit of work. This may achieve the production required but at a reduced productivity. Inversely, too small a crew will result in less being produced per unit of time. Neither condition is desirable. It is the goal of resource allocation to maintain sufficient resources to allow the work to proceed in a smooth or even flow, with a minimum of interruptions.

The estimate (budget) tells us how much labor is needed for a task, but does not tell us when it is needed or in what proportions. For that we need to refer to the schedule. The estimate is also silent about conflicting demands for labor or equipment resources. For example, a single piece of equipment and its crew may be scheduled for two tasks at the same time. Many construction companies have limited resources and are reluctant to hire and invest in training new personnel only to lay them off when they are no longer needed. So to supplement the immediate need, they often subcontract scopes of work. This can offer advantages and disadvantages of varying proportions depending on the application and how the subcontractor is managed. Subcontracting allows the prime contractor to assign the risk via the subcontract and also to fix the cost of the work. With self-performed labor, the cost of which is variable, a measure of control is added. If the crew needs to be increased or has to work longer hours, those are internal decisions. In the case of a subcontractor, it would require their concurrence.

The availability or unavailability of labor, materials, and equipment will influence the manner in which projects are scheduled and thereby managed. Most scheduling methods require that the project manager classify the project as either *time-constrained* or *resource-constrained.*

A time-constrained project has a defined delivery date that is mandated by contract. Time is the constraining factor. If required, resources can be added to meet the delivery date. The delivery date takes precedence over resource availability and it is implied that additional resources will be added to accelerate the work if required. Most everyday construction projects fit the description of a time-constrained project.

A resource-constrained project assumes that the resources available to do the project are limited or cannot be exceeded or over-allocated. Resources are the constraining factor. If production falls behind, the delivery date is extended. This is considered acceptable practice for a resource-constrained project. An example of a resource-constrained project would be a historical preservation, where the artisans or craftsmen required to perform historically accurate aspects of the project are in short supply and cannot be over-allocated on a project.

Another way to define the difference between the two is this: A time-constrained project's delivery date is fixed but the resource levels are flexible. A resource-constrained project has fixed resources, but with a flexible delivery date. The overwhelming majority of construction projects are time-constrained. The same project may have a mix of time- and resource-constrained tasks. It still requires that the project as a whole be defined as time- or resource-constrained. For resource-constrained tasks on a time-constrained project, the project manager must endeavor to keep the resource-constrained task off the critical path of the schedule.

Resource Management

The decision-making process in which tasks are prioritized and scheduled so the utilization of resources is expended efficiently is called *resource management*. While it is extremely difficult to create a perfect utilization of labor and equipment and still meet the critical path of the schedule, much of the lost productivity or wasted time can be eliminated.

Crews that remain constant develop a rhythm or efficiency resulting from repetition and familiarity. If that crew is constantly in flux, it is impossible to develop a rhythm, not to mention the impact on morale. It is extremely counterproductive for skilled labor to be constantly shifted between tasks or moved from project to project. Productivity requires regularity and extended periods of undisturbed work.

Despite every effort by the project manager, there will be times of peak labor demand and other times when there is little work; it is the nature of the business. The demand for a specific resource will fluctuate with the work scheduled for that day. Minimizing the fluctuations of resources on a day-by-day basis is called *resource leveling*. Resource leveling is done by resequencing noncritical tasks within their available float, so that the resource can be available for critical path tasks that have no float. The aim of the resource leveling process is to make the daily utilization of a resource, principally labor, as uniform as possible. This allows patterns of productivity to develop. The scheduling of time-constrained projects is geared toward resource utilization. When demand for a specific resource fluctuates, it presents a management challenge to optimize utilization. Practical remedies to this utilization problem include resource leveling techniques that attempt to balance the need for a resource. The general theory is that resource leveling techniques reschedule noncritical tasks to use their positive float. This reduces peak demands by scheduling the noncritical tasks to absorb the low demand periods for the resource.

The disadvantage of resource leveling is the redirected focus from schedule delivery to resource absorption. Additionally, but maybe more importantly, there is the loss of flexibility caused by reducing the positive float in noncritical tasks. This has the resultant impact of creating more critical or near-critical tasks. In extreme cases of near-perfect utilization, project risk increases as each task becomes critical. This should be avoided.

When a contractor cannot supply the needed personnel to satisfy peak demand and there is no additional labor to contribute, a resource-constrained problem is created. If no recovery plan is instituted, the start of the task is delayed, as is the overall project. The goal now focuses on allocating resources based on priority of need. Concurrent tasks are evaluated based on their physical relationships or overall importance to the task chain. The idea is to minimize the anticipated delay while using the resources available and maintaining the physical relationship of the tasks.

Not every task is a candidate for leveling; however, some are clearly suited to the process. Consider a crane scheduled for the project. It makes perfect sense to reschedule noncritical tasks whenever possible to make use of the crane for the entire day rather than incur the costs of multiple trips to the project if not necessary. It makes good economic sense. Materials do not need to be leveled, but are managed with a different process as discussed in the Material Management section.

Resource Profile

A task's workforce starts small and increases as the need grows. It increases to an ideal or optimal crew, and then as the task reaches completion, the workforce is relocated to another task. If there is no further need for that workforce (e.g., the project is completed), they can be moved to another project or laid off. Figure 12.1 is a graph of a workforce over the life of the task or the project.

The period of peak performance is the center of the graph represented by the relatively flat line. It should be the goal of the project manager to get to that point as quickly as possible, and stay there as long as there is sufficient work remaining in the task.

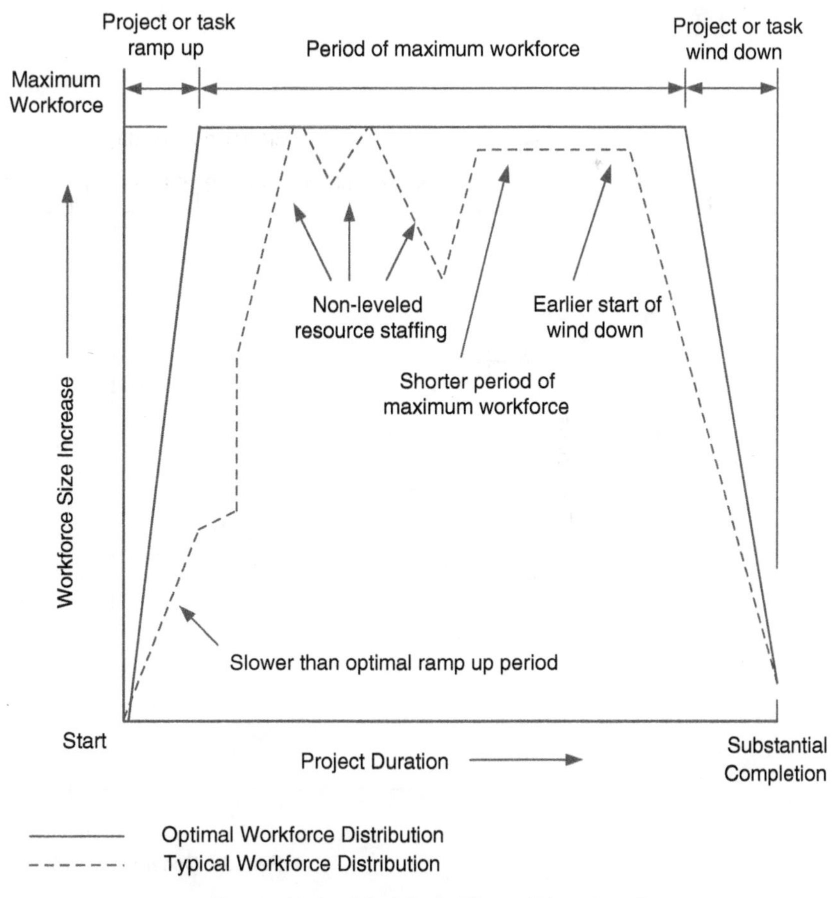

Figure 12.1 Workforce Versus Time Graph

The starting point for resource management is the critical path on the baseline schedule. Analyzing resource needs requires that the project manager be able to create the pattern of labor usage for a particular task. This pattern of resources needs to be graphed as a function of time and is called a *resource profile* graph. It is a day-by-day demand for a particular resource based on a defined schedule for the project. Sometimes called a *histogram*, the resource profile allows the project manager to visualize graphically the needs of a particular resource for the project (Figure 12.2).

The development of the resource profile goes through four distinct steps:

Step 1—Calculate the required quantities of the resources by using the estimate and the schedule together to define the amount of a single resource needed for a given period of time.

Step 2—Distribute the resources to the appropriate crews and activities in the hours they are needed.

Step 3—Summarize the resource usage by time period (day) for the specific resource over the given period being analyzed.

Step 4—Plot the resulting profile of the resource usage per time.

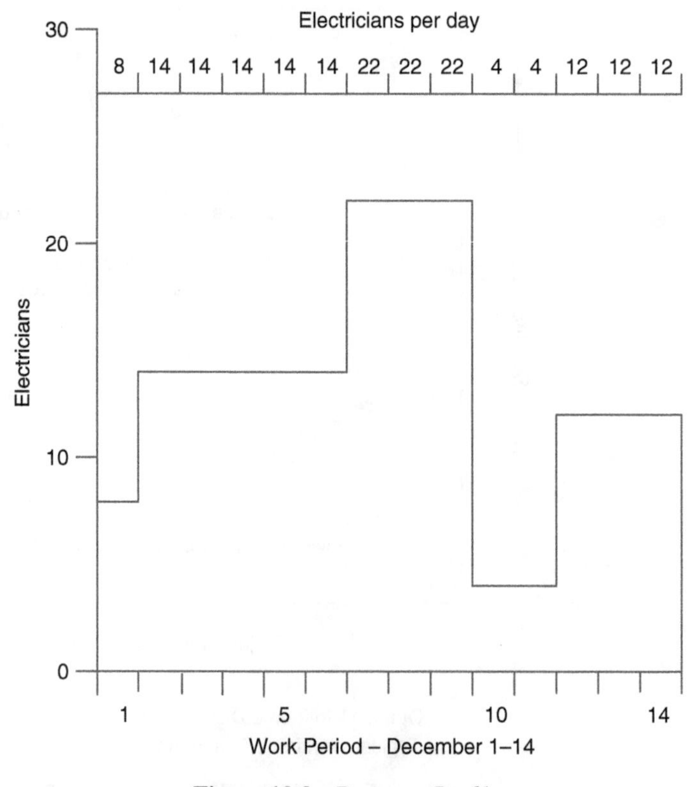

Figure 12.2 Resource Profile

The project manager can then analyze the plotted values and adjust the usage as needed to ensure the continuity of production. This can be done by adjusting some of the noncritical tasks within their available float to abate some of the peak period of resource demand. The author recognizes that project managers are busy people and do not have the time to draw graphs and plot labor usage of every project task. This is not the intent. Histograms are for self-performed work on the critical path. Tasks, whose progress can have a major impact on the project as a whole. This is intended to be the result of a discussion between the PM, superintendent, and foreman for the task.

Splitting Tasks

Another scheduling method to increase resource utilization is called *splitting* tasks. The idea behind splitting an activity is to utilize the resource just enough to accomplish sufficient work on the primary task to avoid delay. The resource is then shifted to the next critical task. When this second task is sufficiently progressed, the resource is then shifted back to the primary task. In reality, this is a fairly common practice that experiences great success, especially if the primary task does not have excessive mobilization or demobilization costs. It also has its greatest applicability where the learning curve is not required to be remastered when the crew returns to the primary task.

Splitting tasks is a common practice with subcontractors. They mobilize the task long enough to get caught up or a little ahead and then off to the next task or even another project. As noted in the previous sections, this jumping from one task to another as a reaction to trying to stay current with the work does not allow productivity patterns to develop, and can reduce productivity, and increase cost. It also has the distinct disadvantage of not being able to predict performance based on the schedule.

Material Management

The careful planning and procurement of material are called *material management*. Material management starts with the correct and timely completion of submittals as well as the tracking of the review/approval process. The project manager must remember that this process must also be monitored and controlled for the subcontractors as well. It requires the coordination of the scheduled need with the material required to perform the work. This includes the research and identification of *lead times*. Lead time is the time between the receipt of approved submittals and/or shop drawings and the delivery of the product to the site. The project manager must define when the material is required and its availability based on that need. If a product is unavailable to meet the schedule, the work must be rescheduled to meet the availability or an alternate product must be researched and proposed as a substitution. Rescheduling of a critical path task may not be an option, especially for tasks that have a physical relationship to the subsequent work.

Proper material management also provides more time for the procurement process. The more immediate the need for the product, the less chance it can be obtained at a reasonable price. Expediting fees for both fabrication and shipping can often result in costs exceeding the budget.

The function of material management starts with the identification of critical items to be purchased in the planning phase and ends with the closeout process when ownership is transferred to the client.

The key objectives of material management are as follows:

- The early identification of critical purchases and lead times
- Timely and correct submission of submittals, samples, shop drawings, and mock-ups for approval
- Tracking and monitoring of all submissions
- Purchase of the item at reasonable or best value pricing, including shipping and storage costs
- Availability of the product in sufficient quantity from a single source
- Coordination of delivery with the scheduled need
- Minimizing of storage costs and rehandling efforts
- Timely closeout documentation: warranties on materials, etc.

While the process may seem rather straightforward, experience has shown that many times materials are overlooked and it takes a Herculean effort to avoid the resulting delay. One effective way to reduce the amount of missed materials is through the development of a submittal log. This requires that the project team review the technical specifications section of the project manual and create a list of the required submittals, shop drawings, samples, and mock-up panels. Once the list is complete, the project manager should prioritize the submittals based on the schedule. While there are some excellent software programs that are set up for the submittal tracking, a simple spreadsheet can be created on Microsoft Excel™ that works well with manual tracking.

It is also a common practice to require subcontractors to identify long lead items. These can be defined as any item that can take more than 30 days to acquire after approval. These key acquisitions of materials or equipment can be identified in the schedule and tracked by both the subcontractor and the prime contractor's project managers. Examples might include steel or joist shop drawings or the submittal for the main distribution electric panel. These are key materials that are clearly critical tasks that can stop a project dead in its tracks.

Material Management in Practice

The actual management of materials can be separated into two competing schools of thought:

1. Deliver the materials just prior to installation. Proponents of this practice claim that it reduces inventory, storage costs, damage due to exposure, and theft/vandalism, and it prevents a premature outlay of funds to pay for stored materials.
2. Provide an adequate supply of materials to be maintained on-site. Special materials/fabrications are ordered and delivered well in advance of need. This requires that they be stored and protected until installed. It also allows special items to be checked for conformance before installation.

Neither is fail-proof and a combination of the two practices is probably the most prudent. Certainly, an argument can be made for stocking some of the more common materials needed so that there is no downtime waiting for materials. However, when storage areas are non-existent or preciously guarded, maintaining an inventory may be impractical. This also can be said for stored products exposed to damaging weather conditions. A counterargument can be made that the chance for theft or vandalism increases with the amount of time a product is stored on-site before installation. Balancing the need and the cost is always the goal. Most contracts provide for the payment of stored materials as shown in AIA G703, a contract document of the American Institute of Architects. On the opposite hand, an undisputed professionalism is displayed when key items arrive with enough time to check them out, and ensure the item is correct, not to mention the added peace of mind.

For tight sites, where storage areas are unavailable, an acceptable option is to rent a storage facility close to the site. Another is to make a deal with local suppliers to keep a reasonable supply of materials on hand for short notice shipments to the site. Both are simple remedies that can have enormous success in avoiding material shortages. Another factor that influences the choice of method is the availability of materials. For remote sites, it may be more practical and more economical to have materials stored on-site.

At the writing of this second edition, the construction industry as a whole is experiencing "supply chain issues." This term is broadly used to define a variety of added costs, shortages, and long lead times for products that in the past were readily available. The project manager is now charged with ensuring everyday products are purchased correctly, in sufficient quantity and a timely manner. The same as always. Some delays caused by material shortages can be excusable and possibly compensable, but the project is still delayed. In addition to identifying long lead items, the project manager can use a Look-Ahead schedule to identify when a product is needed.

Tools for Resource Management

As discussed in Chapter 4 The Schedule, the Look-Ahead schedule is an effective tool for short-term management. This is a micro-schedule or snapshot of the working critical path method (CPM) schedule that looks ahead toward upcoming tasks. The normal period of this tool is 10–15 days ahead, but can be extended. This window can be increased to view tasks a month or more in advance. The intent of the Look-Ahead is to ensure sufficient resources: material, labor, equipment, and (subcontractors') resources are available and ready when needed. This added time can often make the difference between success and failure.

A second effective tool for managing resources is a dedicated review in the weekly meeting. This is as simple as it sounds. Time is set aside in the weekly meeting to discuss and review the status of materials, labor, and equipment for upcoming tasks on the schedule. Knowing the topic will be discussed, proactive project managers can obtain updates on items in advance of the meeting, so delays can be communicated before the last minute.

As always, the submittal log can be instrumental in ensuring products are tracked efficiently. These logs can even be supplemented with required delivery dates for tracking when the product is needed.

Summary and Key Points

The availability and efficient use of resources, especially labor, are always of paramount concern to project managers when trying to control costs. Early attention to the limitations of the available resources and periods of high demand can often be planned for in the scheduling of resources. Even though appropriate durations have been scheduled in the CPM, problems can still arise when demand for the same resource occurs at the same time.

Projects can be classified as either a time-constrained or resource-constrained project. Time-constrained projects focus on the fixed delivery date while resource-constrained projects work around a fixed amount of resources with a more flexible delivery date. While resource management is fairly precise on large labor-intensive projects, its applicability to smaller projects should not be discounted. Resource management is a valuable tool in project control even if the precision is unattainable.

Resource leveling uses the float available in noncritical tasks to absorb the labor available after periods of high demand. While resource leveling is a desirable practice, excessive attention on leveling can refocus the team away from the goal of the scheduled delivery.

Resource management also includes the appropriate and timely management of material and equipment. This starts in the early phases of the construction process with timely and correct submittals. Sluggish supply chains can require a more proactive approach to material management. This can include more regimented review and updates on product status as well as using Look-Ahead schedules as reminders when a product is needed.

Chapter 13 Risk Management will consider the impact of risk on project control and various ways to reduce or allocate risk with other parties.

Key points of this chapter are:

- Project control requires the efficient purchase and timely delivery of products when they are needed.
- Material management can be "just in time" or delivery well in advance, both approaches have benefits and a mix of both is probably the most effective.
- Managing labor resources is often the most challenging, ensuring sufficient, but not too much labor is available when it is needed is a difficult balance.
- It may require that float in noncritical tasks is used to maintain adequate resources when performing critical tasks.
- Project managers should review and explore the tools available to manage resources more closely.

Chapter 12 Resource Management: Questions for Review

1. What are the two goals of resource management?
2. Labor resources can be divided into two types. What are these two types?
3. Material resources can be divided into two types. What are these two types?
4. Scheduling and coordinating resources so that they arrive on site as needed and removed when not needed is called?
5. It is the goal of resource allocation to maintain sufficient resources to allow the work to proceed in a smooth or even flow. True or False?
6. Define the term "time-constrained project."
7. It is not the goal of the resource leveling process to make the daily utilization of labor, as uniform as possible. True or False?
8. What is one disadvantage of resource leveling in construction?
9. What can the project manager expect to see from a resource profile?
10. The actual management of materials can be separated into two competing schools of thought. Explain them.

CHAPTER 13

RISK MANAGEMENT

All projects contain some amount of risk. Construction project risk is defined as the chance of an uncertain event or condition occurring, one that can have a positive or negative effect on the goals of the project. A risk has a cause, and if it occurs, a consequence. Experienced project managers realize it is impossible to eliminate all risk from a project.

Although some risks can have a positive impact, the vast majority, and the main concern of the project manager, are the risks with negative consequences. Some potential risks can be identified in advance and a plan to deal with them can be developed and implemented if they occur. Others can never be anticipated or adequately mitigated. The sources of risk in construction are numerous. Some are within the control of the project team and others, such as extreme weather events, economic changes, and material shortages, are clearly outside of the team's best efforts to control.

It is the goal of a proactive risk management process to identify potential risks, reduce their chance of occurring, and minimize their impact if they do occur. The successful management of project risk can provide better control for the life of the project and can substantially improve the likelihood of achieving the project goals of time and budget.

Components of Risk Management

As previously noted, all construction projects experience some level of risk or negative consequences during their execution. As such, the industry has recognized the need for preemptive measures and has developed a process for managing the risks, both prior to the occurrence and after it happens. This process is called *risk management*. It is the prime objective of the risk management process to identify, analyze, plan for, and manage potential risks. As this text has noted, any event that has been planned for has a far greater chance of being successfully managed.

Risk management attempts to identify a wide variety of potential risks that could occur for a specific project. It is a proactive process to deal with likely and less-likely scenarios. Once the risk has been identified, it can be

Project Control: Integrating Cost and Schedule in Construction, Second Edition. Wayne J. Del Pico.

eliminated or a plan can be developed to mitigate the impact resulting from the risk, should it occur. For the risks that do develop into active events, the goal of the process is to control the ensuing event to completion with as little disruption to the project as possible, while maintaining cost and schedule control over the response process. Clearly, risk management is not without its share of challenges. Risks increase costs and delay the schedule. The goal is to minimize or eliminate both. Any good risk management plan starts with a comprehensive estimate and a realistic schedule.

Risk that is identified early has the best chance of being eliminated through careful planning. Risks that occur early in the process have the best chance of being managed and the results mitigated. This is also true of the cost of the mitigation. The cost impact is less when it occurs early rather than near the end of the project life. Early problems provide greater opportunities to minimize their impact and time to regain control.

Risks can come from a variety of sources: owners, design professionals, contractors, local regulatory agencies, subcontractors, weather, and even the site itself.

Risk Identification

The risk management process is initiated by compiling a list of all potential risks to the project. Ideally, this is done in the planning phase where adjustments to schedule and budget can be made to accommodate the risk. Project team members generate as many risks as possible. Critical thinking is considered positive in this process. They are encouraged to consider even the less-likely risk events. The potential risks can be grouped into *internal* and *external* risks. Internal risks are those typically within the control of the project team and external risks are outside any realistic control process. They can also be separated into technical and management categories. The risk identification process can be expanded beyond the immediate team to other stakeholders whose performance, or lack of performance, can impact the success of the project.

Team members focus on the risk events that produce the negative consequences, instead of the overall broad scope of events. The Work Breakdown Structure (WBS) is used as a guideline to make sure nothing is overlooked. Potential risks are considered for macro levels of the WBS and then drilled down to the micro level tasks.

Another tool of the risk management team is the *risk profile*. A risk profile is a list of questions that have been classified in traditional areas of known construction risk. Risk profiles are updated and refined continually based on prior events from past projects. The historical management experience of the company determines the questions. Risk profiles acknowledge the unique management characteristics of the company and the managers. They can also consider the financial reserves of the company for dealing with any risk event. A risk profile is meant to challenge the assumptions that may have been made at bid time.

Risk Assessment

Once the list of potential risks has been identified, the next step is to assess both the likelihood of the event and the severity of the impact. Some of the risks can be discounted as insignificant, while others require a more in-depth evaluation. The project team will require a method of qualifying the risks as inconsequential or significant. The significant risks will be classified based on the probability and range of the impact.

The most frequently employed method of rating the significance of the risk is through the *scenario analysis*. The scenario analysis classifies the significance of the risk in terms of the likelihood or probability that the risk will develop into an event, and the severity of the impact if the risk does develop into an event.

It is a recognized fact that there is an important distinction between the probability and impact factors of a risk. Impact is generally considered a more detrimental factor than probability. As an illustration of this point, a 10 percent chance of losing $100,000 is a more serious threat than a 90 percent chance of losing $1,000. As a result, there needs to be a way to rate both probability and impact independently for a task. Considering all of the threats to a successful project that a team could develop in the identification phase, there must be a methodology for rating risks according to severity. The *impact scale* provides just such a rating system. It allows the project manager to prioritize what risks are in need of attention. The impact scale employs a numerical value ranging from a very low (1) likelihood to a very high (5) likelihood. The same applies to the impact: very low (1) impact to very high (5) impact. This is done for each significant risk to cost and schedule goals. Table 13.1 is an example of an impact scale for a fixed price contract.

The impact scale is developed on a project-by-project basis with varying delineation. Additional goals, beyond cost and schedule, may be added depending on the type of contract delivery or the nature of the risk. The impact scale can then be used as the basis for a comparison that allows the

Table 13.1 Impact Scale for a Fixed Price Contract

	Numerical Impact Scale				
	1	*2*	*3*	*4*	*5*
Category	**Very low**	**Low**	**Moderate**	**High**	**Very high**
Cost	Insignificant cost increase	< 10% cost increase	10–19% cost increase	20–49% cost increase	Greater that 50% increase in cost
Time	Insignificant time increase	< 5% time increase	5–9% time increase	10–19% time increase	Greater that 20% increase in time

project manager to assemble and rate the risk for a task based on a numerical scale. This is called a *risk severity matrix*. The matrix is organized around the probability and impact of the risk event. More weight is given numerically to an event that has a high likelihood and high impact than one with low values for both. The matrix can be color-coded to reflect the severity of the event. It is divided into red, yellow, and green zones, representing severe, moderate, or minor threats, respectively. The lower left corner is the green range which represents low likelihood, low impact. The yellow range extends down the center of the matrix and identifies risks that expose the project to a moderate likelihood and moderate impact. The red zone is the top right corner and is for risks with high likelihood and high impact. The use of the traffic light colors conveys the importance of the risk with a recognized color scheme. Figure 13.1 is an example of a risk assessment matrix for identified risks to project schedule and cost.

The risk severity matrix provides a basis for deciding on which risks require further evaluation and a possible response plan. The red zones receive the first priority, with yellow zones next. Green zones may be considered a less than serious consideration and are marginalized, or even ignored.

Another consideration in the assessment process is how easily the risk can be detected. In addition to the probability and impact, the team must

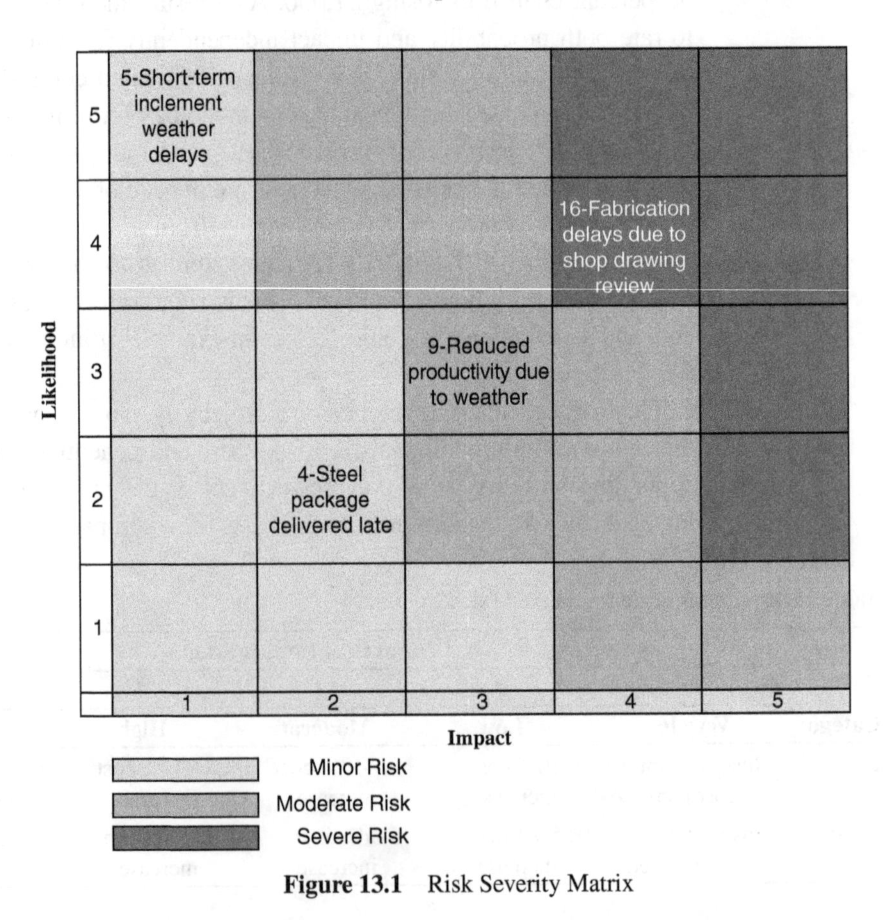

Figure 13.1 Risk Severity Matrix

consider how ease of detection can affect the risk. Detection can be defined as how easily the risk can be discovered as imminent. Risks that are easily detectible pose less of a complication to the threat than those that are difficult to detect. To analyze the ease of detection for a threat, an impact scale for detection is created. The scale rates easy detection with a value of 1, and risks with a high difficulty of detection at 5. Differing latent subsurface conditions such as unsuitable soils would be on the higher side of the scale if the project came without a subsurface investigation report. A labor strike might be on the lower end of the scale as they rarely occur without warning.

To use this added dimension requires the emergence of another formula. The formula is called the *Failure Mode and Effects Analysis* (FMEA), and includes the detection into the factoring of risk severity:

$$\text{Probability} \times \text{Impact} \times \text{Detection} = \text{Risk Severity Value}$$

In this formula, each of these three factors is rated on a 1–5 scale with 1 as the lowest and 5 as the highest. The risk can be prioritized based on the overall score. Consider Risk A and Risk B in the following example where the value of each factor of the impact scale is as follows;

$$\text{Risk A} = 2 \times 2 \times 4 = 16$$

$$\text{Risk B} = 2 \times 5 \times 1 = 10$$

By this simple example one would surmise, based on the mathematical calculation alone, that Risk A posed the greater threat to the project cost or schedule. This demonstrates the flaw of using the numerical value of the FMEA without common sense or a further review. While Risk A is extremely difficult to detect, its probability and impact are low. Risk B, by comparison, presents a formidable impact should it occur, even with a low probability. Risk B would require further attention due to the rating of 5 on the impact scale. The reader must recognize that the FMEA score is only the opening gambit and requires further investigation.

Risk Response Plan

When a risk event has been identified, assessed, and discussed, the next step is to decide on a plan for responding to the event should it actually occur. A response plan must be specific to the event and can be managed in five ways. The decision of the project team will be focused on which method of risk response provides the least impact in terms of cost and schedule: mitigating, avoiding, transferring, sharing, or retaining.

Mitigating Risk

Ideally, the best method to deal with risk is by its elimination. If that is not possible or practical, a second option is to reduce the likelihood that the event will occur and the impact of the event if it does. This is called *mitigating risk* and is traditionally the first option of the project team. The focus of mitigation is on reducing the probability that the event will occur since this may actually render a plan to reduce its impact unnecessary. A good starting point is to determine the cause of the risk. Understanding the cause can often be critical to mitigation.

The next best technique is to reduce the impact of the event if it occurs. This strategy focuses on which factors drive the cost up and delay the schedule. Providing alternatives to the process, product, or entity driving the risk event helps to reduce its impact. A simple example is having multiple suppliers for lumber should one supplier be unable to fill the need in a timely or cost-effective manner.

Avoiding Risk

Avoiding the risk altogether is similar to eliminating it. Avoidance may require a rethinking of the project plan. Changing specific products and processes to ones that are tried and true can help avoid risk. Avoiding risk is often limited to those within the control of the project team. The ultimate avoidance of risk is not to bid the project. While this may seem extreme, every estimator has reviewed bid documents that are incomplete, contradictory, and for all intents and purposes defective, yet still put out to bid. While it is acknowledged that there is always some contractor that will bid the work, sometimes it is best to decline. This is typically the result of an upper management decision not to expose the company to the risk or the inevitable legal harangue that will follow. Sometimes the best way to avoid the risk is to avoid the project.

Transferring Risk

One of the most common response plans is to transfer risk to another party. The risk still exists; however, it is the responsibility of others. This is essentially what occurs when a prime contractor subcontracts a portion of work to a subcontractor. The key to transferring risk is that the receiving party is in a superior position to deal with the risk. This superior position, for example, can be better training, experience, or the most technologically advanced equipment. The simple act of transferring the risk may in and of itself improve the chance to mitigate the risk. The important consideration is to ensure that the receiving party is capable of assuming the risk. Risk can also be transferred through exculpatory language in the contract. There are many examples of exculpatory language meant to transfer risk such as the *No-Damages-for-Delay* clause that makes the contractor's sole remedy for delay an extension of time. Other similar contract language may have disclaimers on the plans or in the specifications shifting the risk to the contractor. Examples that come to mind include the responsibility to verify the dimensions in the field, or the standard disclaimer that avows any responsibility for the soils reports. The key to transferring the risk is that the receiving party be experienced in managing or mitigating the risk.

Sharing Risk

The act of sharing risk involves multiple parties assuming different responsibilities in relation to the risk event. Sharing risk is intended to ensure that the unique talents of each party prevail in managing the risk. The joint venture is a classic example of sharing risk between two (or more) companies. Two contractors combine talents to better execute the project and, in turn, share the risk. The success of the joint venture lies in defining the responsibilities of each party. On some projects, the risk is shared between the owner and the contractor through special contract provisions. As a result, the cost to perform the work can be less since the owner shares the burden of risk with the contractor in lieu of the traditional relationship. In most traditional delivery methods, a disproportionate amount of the risk is retained by the contractor. In the contract relationship, called *partnering*, each party to the contract is allowed some flexibility in cost and schedule parameters in order to bring the best value to the project. The partnering relationship encourages improved performance based upon the expertise that each party brings to the project.

Retaining Risk

In some circumstances, the risk impact is too large to transfer or share within reasonable economic terms. The risk may also be better suited to stay with the originating party for its ability to detect and control the risk. In this situation the risk is retained especially if the probability of the risk is miniscule. The risk is contained or managed by developing a way in which to handle the event, should it materialize. In retaining the risk, the party responsible for the risk develops a plan to manage the risk so that time is not wasted while a plan is developed. The sources creating the potential risk are monitored closely so the response can be immediate, with minimum delay to the schedule.

The more time spent on risk response in the planning phase, the more successful a team can be in handling the risk, should an event occur.

Risk Response Control

The final step in the risk management process is risk response control. Risk response control is implementing the risk response strategy developed by the team. How will this event be managed? The control process requires that the impending risk event be carefully monitored for threshold events that would warrant the start of the response. Once the response has been set in motion, it must be monitored to assure it is having the intended effect on the risk. The response may require minor tweaking to arrive at the desired outcome or may require that another plan be developed and implemented.

Risk response control is analogous to the progress tracking that goes on in the project control cycle. The status of the risk under control must be updated regularly at meetings and in reports until the event has been eliminated or mitigated, or the period of concern has passed. The team must be vigilant for new unanticipated risks that may develop, especially on large and complex projects. This is also true for projects of long duration (years) where complacency and management fatigue can set in. For projects of this nature, the risk identification, assessment, and control cycle is repeated regularly until completion.

Another focus of risk control is clear assignment of responsibility for the team members. Identified risks that will require a response must be assigned to a team member. Clear assignments of responsibility go a long way toward eliminating "I thought you were handling that!" statements that accompany missed opportunities to manage a risk. The assignments must be made by mutual agreement and not assumed by default. If the management of the risk is not formally assigned and accepted, human nature has shown that the risk will be ignored.

The risk management cycle must be an ongoing process from the planning stage to the closeout. Team members must keep one eye on identified risk and one eye on potential new threats. As new threats develop, their potential probability and impact must be assessed and managed to completion.

Risk Contingency Planning

It is often necessary for the project team to create a *risk contingency plan* to deal with the impending threat should it become a reality. The contingency plan is intended to reduce or neutralize the unwanted or negative impact of the risk. Like all planning, the more thorough the investigation, the more accurate and effective the planned response can be. Failing to create and agree upon a contingency plan will delay the implementation of a response. This can, and often does, add to the magnitude of the impact. Lack of a plan when the risk event occurs will lead to the adoption of a poorly thought-out risk response. Contingency planning well in advance of the occurrence allows time for review and evaluation of the proposed response, not to mention an estimate of its cost. It allows sufficient time for the best option to surface and its smooth implementation.

Thresholds for triggering the implementation of the plan must be spelled out clearly and accepted prior to the occurrence of the event. This prevents debates when action is what's needed. Each party to the plan must know what their level of *functional authority* is and when it can be executed. Functional authority can be simply defined as; "what decisions can I make on my own, and what decisions require concurrence of senior management or the team?" Experienced project managers establish protocols for enacting the response plan well in advance of the need. They also are clear as to their level of functional authority.

The contingency plan by definition disrupts the normal flow of work and, as a result, has a cost and schedule impact, both of which must be planned for in advance. It may also require the use of resources that were previously committed to contract work. Specifically, how will the delay caused by the response be accommodated in the schedule and where will the funding to pay for the response come from?

Contingency Plan Funding

For all risk contingency plans, there must be a source of funding to pay for the response and a "work-around" or rescheduling to accommodate the delay. For many responses, the work-around in the schedule is often part of the cost estimate for the response. In short, the response to the delay is some form of acceleration that increases cost. For a risk caused by the prime contractor in a stipulated sum contract, the funding may come from the profit column. The funding can also be sourced from the difference between the estimated profit and the target profit. It can also be funded from the savings due to skillful management of the project to date.

For owner-caused risk events the funding may be their responsibility. In this case, the source of the funding may be a special reserve fund created for this purpose. This is a common practice among savvy owners, especially for unique projects.

Regardless of the ownership of the cost, the source of funding must be known and agreed to in advance of the need. To fail to establish a source of funding can render the plan moot. Contractors or owners who refuse to establish contingency funds until the risk has occurred are being unrealistic. To make the assumption that if the money is available, it will be spent does not recognize the nature of construction. Safeguards or check-and-balances can be established so that the contingency fund does not become a remedy for poor management. Reserves for risks that have been identified, estimated for, and deemed likely to occur have established *budget reserves*. Budget reserves are funds set aside for what appears to be the inevitable. They should be established on a risk-by-risk basis and should not be estimated as a percentage of the project. These budget reserves actually comprise a portion of the Management Reserve discussed in Chapter 8 Analyzing and Reporting Variances in Schedule and Cost. The unknown and highly unanticipated risks that materialize into full-blown events need to have a contingency as well. Contingencies for these types of events are also added to the management reserves totals. Management reserves are often set aside based on a percentage of the total project cost. Percentages can range from 2–3 percent for simple new construction to a whopping 100 percent for complex historical restorations. The agreed-upon percentage is often supported by historical cost and schedule data and the uniqueness of the project. Projects with incomplete documents also rate a higher percentage of management reserve.

Occasionally designs are released for bid before they are fully complete with the idea that the contractor will find what is missing or in error during the bid process. The design professional will then correct or fill in the missing information by addenda. This is called a *minimally viable design* or *MVD* and is intended to save design fees. It is supplemented with a larger management reserve. This is a dangerous practice and can often result in delays and overruns caused by the volume of changes that occur, that may have been missed during the bid process. There is no substitute for a well-developed and professionally reviewed set of documents as a means of reducing contingency funding.

Probability Analysis in Scheduling

A risk response can have a cost impact associated with it, but it can also have a schedule impact. Most professionals acknowledge that the main use of the critical path schedule is for project control. More to the point, the management of risk may be encountered during the control process. There is an analytical technique available to the project manager in which risk's impact on the schedule can be analyzed. It is intended for critical path tasks with potentially high schedule risk. This technique is called *Program Evaluation and Review Technique* (PERT), and can be used in the planning phase to evaluate the duration of specific tasks on the schedule. Adjusting the time of the critical path task through this kind of probability analysis can reduce risk. It can also be used to determine the schedule risk of the project as a whole, based on a likelihood of outcome projection. PERT assumes a statistical distribution based on optimistic and pessimistic projections. It can then be used to predict the most realistic duration for a task.

PERT simulation is not unique to the construction industry; it has broad appeal in many industries. Used in conjunction with other previously mentioned techniques, it can be valuable in analyzing the full impact of risks on specific critical path tasks. High-risk tasks that lie on the critical path have zero float and are at greater threat of failure. To reduce this threat and improve the chances of success on the high-risk tasks, the project manager analyzes the likelihood or probability that the subject task will complete in the specified duration.

PERT was formulated by the Department of Defense in the late 1950s during the development of the Polaris nuclear submarine. Having no previous historical data to use as a guideline for determining the project timeline, the Navy developed PERT as an aid in predicting duration. PERT is a statistical analysis of tasks along the critical path to determine the probability of a task or the project in the prescribed duration. PERT is not without its flaws since it ignores the impact of near-critical path tasks. However, limitations notwithstanding, it adds some distinct value to the control process.

There are three key pieces of information necessary to start a PERT analysis:

- The most likely duration for completing the task.
- The most optimistic duration for completing the task if the work proceeds ideally.
- The most pessimistic duration for completing the task if the work experiences delay as a result of the task experiencing difficulty.

With this information, the PERT formula can be used to calculate the $PERT_{Mean}$, which is a “weighted average” since the most likely time estimate is weighted four times as much as the optimistic and pessimistic values. The formula for the $PERT_{Mean}$ is as follows:

$$PERT_{Mean} = \frac{O + 4M + P}{6}$$

where

O = the most optimistic duration for completing the task if the work proceeds ideally

M = the most likely duration for completing the task given the normal course of events

P = the most pessimistic duration for completing the task if the work experienced delays as a result of things going wrong

The $PERT_{Mean}$ is affected slightly by either the optimistic or pessimistic value depending on which one is farthest from the most likely value. This results in moving the $PERT_{Mean}$ toward the value farthest from the most likely value M. It should be noted that the project manager should develop a list of criteria that is the basis of the things that could go wrong and raise the pessimistic value P.

The next step in the PERT analysis is the calculation of the $PERT_{Standard\ Deviation,}$ which is defined as the variability of the calculation of each task. It is determined by subtracting the optimistic value from the pessimistic value and then dividing by six:

$$PERT_{Standard\ Deviation} = \frac{P - O}{6}$$

where

O = the most optimistic duration for completing the task if the work proceeds ideally

P = the most pessimistic duration for completing the task if the work experienced delays as a result of things going wrong

A large $PERT_{Standard\ Deviation}$ is viewed as a wide range between the optimistic and pessimistic estimates. If the standard deviation is small, it means the optimistic and pessimistic durations would be relatively close in range. The larger the $PERT_{Standard\ Deviation,}$ the less confidence there is in the estimate of the duration. The standard deviation's size can then be used in comparing the relative risk one task has to another. Here is an example of an analysis for Task A in which the $PERT_{Mean}$ and $PERT_{Standard\ Deviation}$ are calculated in terms of workdays for the case where O =10 days, M = 12 days, and P = 16 days:

$$PERT_{Mean\ A} = \frac{O + 4M + P}{6} = \frac{10 + 48 + 16}{6} = 12.33 \text{ days} \approx 13 \text{ days}$$

$$PERT_{Standard\ Deviation\ A} = \frac{P - O}{6} = \frac{16 - 10}{6} = 1 \text{ day}$$

The values for $PERT_{Mean}$ and $PERT_{Standard\ Deviation}$ can then be used to calculate the likelihood that Task A will finish within 13 days.

The formulas for calculating the probability are derived from a statistical analysis that is intended to help improve the decision process in assigning the duration of a high-risk task.

The previous results can be used with the *normal distribution* that forms the basis of statistical distribution theory. Normal distribution, also called the *Gaussian distribution*, represented by the bell-shaped curve, is the work of Carl F. Gauss, an early nineteenth-century German mathematician. The bell-shaped curve can be used to describe physical events such as task duration on a construction schedule (Figure 13.2).

According to Gauss, the mean represents a 50-50 chance. He also noted that 68 percent of the data he collected fell within one standard deviation

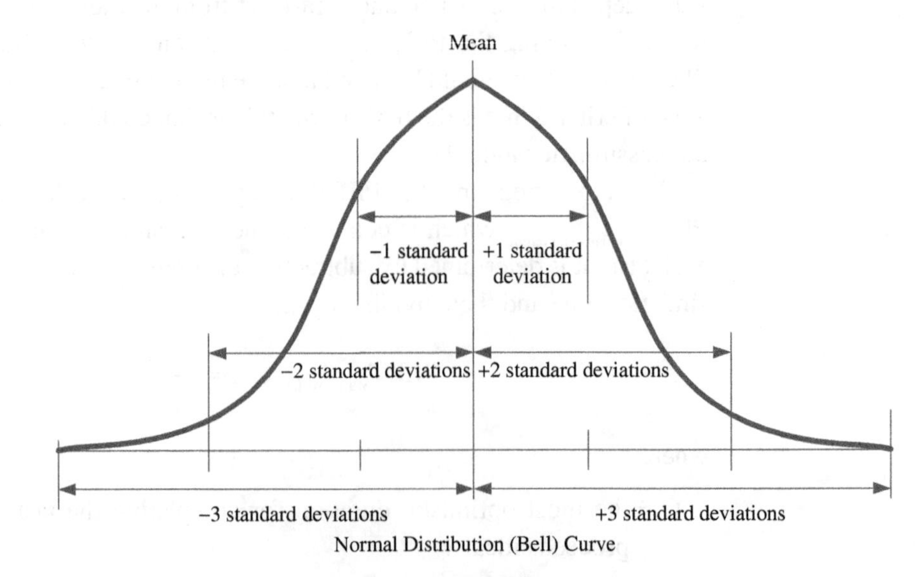

Figure 13.2 Bell-Shaped Curve

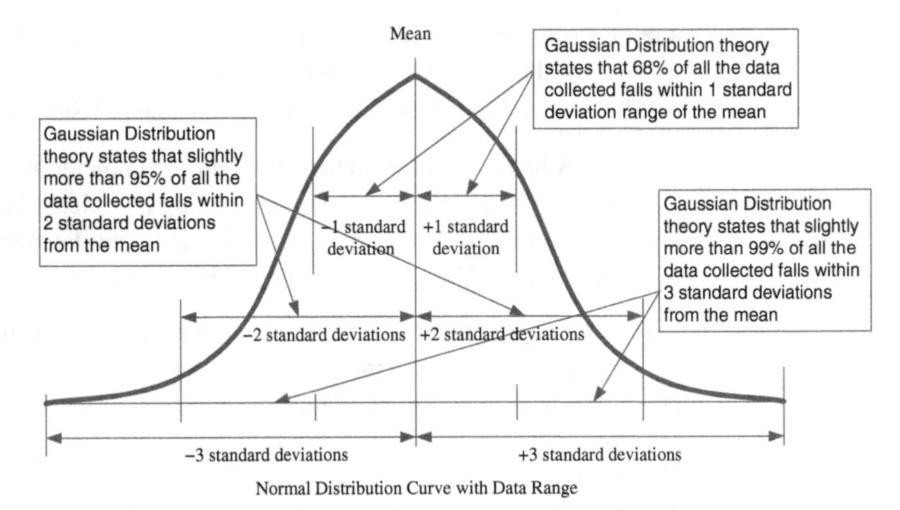
Mean
Gaussian Distribution theory states that 68% of all the data collected falls within 1 standard deviation range of the mean
Gaussian Distribution theory states that slightly more than 95% of all the data collected falls within 2 standard deviations from the mean
–1 standard deviation
+1 standard deviation
Gaussian Distribution theory states that slightly more than 99% of all the data collected falls within 3 standard deviations from the mean
–2 standard deviations
+2 standard deviations
–3 standard deviations
+3 standard deviations
Normal Distribution Curve with Data Range

Figure 13.3 Range of Data on a Bell-Shaped Curve

of the mean (Figure 13.3). Applying Gauss' theory to the values for Task A noted before, it can be concluded that there is a 68 percent chance that the duration of Task A will fall within plus or minus one day of the $PERT_{Mean}$ A of 13 days. If the mean represents a 50-50 chance of finishing on that date (day 13), then it can be concluded that half of the 68 percent will fall on either side of the mean. This results in an 84 percent chance (50 percent plus 34 percent) that Task A will finish in 14 days. Conversely, there is a 16 percent chance (50 percent minus 34 percent) that Task A will finish in 12 days.

Gauss went on to calculate the amount of data that was inclusive in the bell-shaped curve based on multiples of the standard deviation. He concluded that slightly more than 95 percent (95.4 percent) of the data collected falls within the range of two standard deviations, and that a little more than 99 percent (99.6 percent) of the data collected falls within the range of three standard deviations.

If this is applied to the Task A example, then it could be stated that the probability for Task A being completed in 15 days (13 days + two standard deviations of one day each) is slightly more than 97.5 percent (95% ÷ 2 = 47.5% + 50%). The actual number Gauss used is 95.4 percent which, divided in two (47.7 percent) and added to the mean, is 97.7 percent. If we deduct 47.7 percent from the mean of 50 percent, it results in a 2.3 percent probability that Task A will finish in 11 days (13 days minus 2 days standard deviation).

If the range for three standard deviations is calculated, a little more than 99 percent (actually 99.6 percent) of the data collected falls within the curve. With half of the 99.6 percent on either side of the mean, then there is a 99.8 percent chance (99.6 percent ÷ 2 + 50 percent) that Task A will finish in 16 days (13 days + 3 days standard deviation). On the other side of the mean, there is a -0.2 percent chance (99.6 percent ÷ 2 minus 50 percent) that Task A will finish in 10 days.

What does all this mean for the project manager? Simply stated:

- Task A has a 50 percent chance of being completed in 13 days.
- Task A has an 84 percent chance of being completed in 14 days.
- Task A has a 97.7 percent chance of being completed in 15 days.
- Task A has a 99.8 percent chance of being completed in 16 days.

The project manager and the project team can decide if the potential reduction in risk is worth the extra day(s) to the schedule's critical path. This decision is often based on the construction company's risk tolerance.

Merge Bias

One of the principal objections to the use of PERT for calculating the probability of the entire project is that an analysis technique can only be used on a single path through the schedule. The summation rule of statistics assumes that data under analysis is "mutually exclusive," which is understood to mean that no other task in the schedule will impact the analysis. Project managers know that this is not the case in construction. Activities that merge to finish at a milestone can have an impact on concurrent and succeeding activities. It is also routine for multiple critical paths to exist in a project as well as near critical or *subcritical* paths. Subcritical paths are distinct independent strings of activities with small total float. As float is reduced on these subcritical path tasks, the probability of completing the project on time is reduced because the likelihood that the subcritical path will cause delay increases.

The concept of *PERT merge bias* can be defined as the probability that delay to the schedule increases significantly in proportion to the number of critical or subcritical paths with minimal float. The PERT analysis process ignores the lower tier critical paths or the subcritical paths in the calculations. As a result, the risk is underestimated for convergence points of multiple activities.

PERT can calculate the probability of any number of paths independently; however, rarely does a project schedule consist of one critical path. At points where two or more paths converge, a statistical sum is required to represent the outcome probabilistically. The statistical sum considers the impact one path has on another. The formula for a statistical sum in the case of $Path_A$ and $Path_B$ is

$$P(Path_A + Path_B) = P(Path_A) \times (Path_B)$$

where

P = Probability in percentage ($x/100$)

$Path_A$ = Critical or near critical $Path_A$ converging with $Path_B$

$Path_B$ = Critical or near critical $Path_B$ converging with $Path_A$

To illustrate the point, assume that $Path_A$ has a 50 percent chance that it will finish on the milestone date. Now assume that $Path_B$ has an 84 percent chance (this assumes $Path_B$ has float, which is equivalent to one standard deviation). To calculate the likelihood that both of them will reach the milestone together, multiply the percentages:

$$0.50 \times 0.84 = .42$$

There is a 42 percent that they will finish on the same milestone date. Now multiply this with another merging task that has a float equal to two standard deviations (97.7%):

$$0.50 \times 0.84 \times .977 = .41 \text{ or } 41\%$$

The percentages decay even further as additional paths are added. This illustrates merge bias and shows why many experts warn against its use as a schedule risk analysis technique. Despite the limitations caused by merge bias, performing a PERT analysis for high-risk tasks or the entire project is a step in the right direction. It is a step toward improving the chances of success for the project.

Monte Carlo Simulation

Another more widely accepted method of calculating the probability of project success is to create a simulation of multiple possible scenarios. This method is called the *Monte Carlo simulation* after the city in Monaco famous for its games of chance. It is a probabilistic approach to determining the confidence levels in finishing tasks on time using a statistical distribution. The Monte Carlo simulation emulates the random outcomes common to games of chance by randomly selecting variable values to simulate real life. This exercise provides a multitude of scenarios that can be used to address schedule risk. It randomly assigns durations from predetermined values for optimistic, most likely, and pessimistic durations similar to the PERT analysis. The expected statistical distribution in the Monte Carlo simulation, however, uses the *beta distribution* instead of a bell curve. The beta distribution differs from the bell-shaped curve in that it favors increased distribution to the right of the mean (Figure 13.4).

The beta distribution is said to better represent tasks in a schedule since tasks that improve their scheduled duration do so in small increments, whereas tasks that slip their schedule often do so by larger increments.

Monte Carlo simulations are performed by computer and are not practical to execute by hand, due to the sheer volume of calculations. For each sequence, the software assigns durations randomly from the optimistic, most likely, and pessimistic estimates of duration provided for each task. The software applies the distribution and runs calculations for each task in the CPM. As a result of each sequence run, the end date of the project is tabulated, then expressed as a probable outcome. The results are typically listed in 5 percent

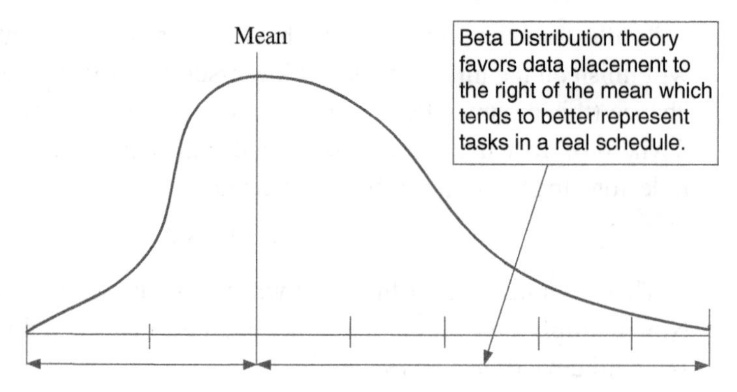
Mean
Beta Distribution theory favors data placement to the right of the mean which tends to better represent tasks in a real schedule.

Figure 13.4 Beta Distribution Curve

Table 13.2 Trial Results for Monte Carlo Simulation

Probability Table for Project Completion			
Probability	**Substantial Completion**	**Probability**	**Substantial Completion**
5%	12/10/2022	55%	1/3/2023
10%	12/12/2022	60%	1/5/2023
15%	12/16/2022	65%	1/7/2023
20%	12/17/2022	70%	1/10/2023
25%	12/19/2022	75%	1/12/2023
30%	12/21/2022	80%	1/13/2023
35%	12/23/2022	85%	1/14/2023
40%	12/26/2022	90%	1/16/2023
45%	12/28/2022	95%	1/18/2023
50%	12/29/2022	100%	1/19/2023

increments from 5 percent to 100 percent, each defining the *confidence level* in the project's finish by a specific date. The confidence level is a measure of the statistical reliability of the predicted outcome, in this case project completion. Table 13.2 is an illustration of the expected trial results.

The results of Table 13.2 allow the project team to select a completion date based on the confidence percentage that provides the desired comfort level.

Monte Carlo simulations and the calculation of PERT values, including merge bias, are done exclusively by computer. To do the calculations by hand would be time-consuming and subject to human error. However, it is essential to recognizing the potential risks and the reduction of those risks to understand what the software is actually doing. The above explanations were to educate the reader as to the benefits of these calculations.

Contract Type and Risk

For all projects, the type of delivery contract will impact risk and project control procedures in general. There are several types of contracts that are common in construction. A brief review of each and the controls and risks involved is now given.

Stipulated Sum

One of the most common contract types is *stipulated sum*, also called firm fixed price, lump sum, or single prime, which is a traditional form of delivery. It is based on the design-bid-build model in which a contractor bids on and is awarded a contract based on the scope identified in the plans and specifications. It is a fixed price based on the documents and poses a disproportionate amount of risk on the contractor. Out-of-scope work requires a change order. It does, however, offer the professional contractor the opportunity to improve their financial position through skill and experience. It is almost always a "closed-book" contract that does not allow the owner to view the costs accumulated by the contractor in the performance of the work. Contracts of this type require close scrutiny and control of cost and schedule. A risk management analysis and plan are essential for success as project control is in the hands of the contractor and there are no remedies for poor management, incorrect estimating, or flawed execution of the work. There are frequently penalties for contractor-caused delay.

Cost of Work Plus Fee

Work performed without a maximum price is called *Cost of the Work plus Fee*. It is also commonly called time and material or stock and labor. An estimate is not required since scope is not fully defined. It is typically for limited scope or where the full design has yet to be determined. It is frequently used in emergency work. It requires that the terms *work* and *fee* be defined in detail so there is no confusion. This contract type poses little if any risk to the contractor and project control is more of an accounting task. It provides maximum flexibility for the owner along with the majority of the risk. The contract is "open book" where the owner is allowed access to the contractor's accounting and costs are supported with invoices, time cards, rental slips, etc. Profit is limited by the established fee and there is little if any incentive to manage costs, since the fee is typically a percentage of cost. Schedule is unknown due to lack of defined work scope.

Cost of Work Plus Fee with a GMP

Another version of cost plus contract is called *Cost of the Work plus Fee with a GMP*. The GMP is *guaranteed maximum price* or cap on the contract amount. It is a blend of the stipulated sum and cost of the work plus fee contracts. It is a contract model that shares risk. It is based on documents that are often incomplete with allowances filling in for missing design. The contractor provides an estimate and schedule for the scope identified at bid time. Both cost and schedule can be negotiated. The contractor's fee is identified and projects are often supplemented with a contingency fund. Rules for use of the contingency fund are well established and require concurrence by the owner. This type allows flexibility, but requires a change order for out-of-scope work, similar to the stipulated sum contract. There can be penalties for inexcusable delay. The contract is "open book" where the owner is allowed access to the contractor's accounting and costs are supported with invoices, time cards, rental slips, etc. The contract sum is guaranteed, but not individual cost accounts, so a cost overrun on one account can be supplemented by an underrun on another account. The contractor's financial position can be improved through incentives gained by sharing cost savings. It requires detailed cost accounting and solid project controls as both the GMP and schedule are typically guaranteed, since there are damages for inexcusable delays.

Design Build

An alternative to the design-bid-build methodology is the *Design Build* delivery system. Both design and construction are under one contract. The criterion for design is not flexible but the way to achieve it is. The design is often based on cost-effectiveness. The design may mature as the work progresses. Out-of-scope work requires a change order. Costs and schedule are typically guaranteed by the contract. This delivery type can often reduce risk due to the flexibility of the design options. It does require detailed project control procedures and may even be an "open-book" type contract with allowances for those scopes yet to be determined. This type of contract has a maximum price similar to a stipulated sum contract. Risk assessments and plans include input from the design team. Risk can often come from the design portion which is not part of the typical risk palette of the contractor.

Construction Management

Under this methodology the owner secures the services of a construction management firm to coordinate and manage the project from inception. Construction management has two different levels of service. The Agent model acts to manage the cost and schedule of the project but without penalties to the fee for failure. The At Risk model manages cost and schedule and the fee is at risk for cost and schedule overruns. The CM can also be the constructor along with other multiple prime contracts. The CM is responsible for project controls and the development of risk management techniques and responses. The contract is "open book" and subject to scrutiny by the owner. As with other delivery systems, there is a cap on costs and schedule at commencement of the work. In the organizational chart the CM is placed between the owner and the design team and contractor.

Partnering

A fairly new contract model requires that the owner, design professionals, and contractors set some common objectives and goals for the project before commencement. All parties are tasked with working to achieve the goals within a schedule and budget. The contract between the parties is "open book" and is geared toward cost-effectiveness with no party earning too much profit. It is often used on multi-phase projects with long durations. It is also intended to reduce the risk of legal disputes from claims and to create an atmosphere of trust and communication not typical in the construction process. There are numerous versions of this delivery method but most have flexibility in the budget and schedule. Savings through value engineering and cost and schedule control can be shared among the parties. All parties are tasked with developing and executing cost controls for their work, and the project as a whole. Regular meetings reviewing progress and cost are essential to status reporting. A large part of the partnering process includes risk assessment and plans to respond to the risk. It is intended to reduce risk between the parties through openness.

This is by no means a complete list of delivery methods. Others include Fast-Track, Multiple Prime, and Turnkey, to name a few more. The selection of the contract type is typically done by the owner, often with guidance from counsel and/or the design professional. The contract delivery method is intended to reduce the risk to the owner. Each contract type can include exculpatory language to further reduce the risk of the owner. Quality and completeness of the documents go a long way to enhancing the effectiveness of the agreement.

Contract Modifications and Impact on Risk

The majority of contract delivery methods provide that an owner has a unilateral right to make changes throughout the project. Their only requirements is to pay for the change and extend the time for performance, if warranted. As many seasoned project managers will attest, many owners wait until the latest possible time in which to exercise this right. Many project managers have experienced the bevy of change orders that are authorized in the eleventh hour when owners see their window of opportunity to get the work done closing. This late-stage flurry of change orders can often have a previously unrecognized risk component when change orders are vying for the same limited resources and remaining schedule. It is wise for the project manager to preserve the rights of the contracting team by reviewing the risk component of the combined change orders prior to accepting them. Also an expiry date on individual change orders or language stating that the contractor reserves the right to review and revise the change proposal at the time it is awarded may also be helpful. This is especially true if the conditions have changed from the period in which the change order was priced. Proposed changes priced early in the process often must be revisited and reviewed for impact on cost and schedule in real time. Proposed changes that are not accepted till later in the project can have an adverse impact on other scopes of work that were not an issue when the work was priced and the change proposal submitted. This goes beyond the in-sequence and out-of-sequence argument. It is often advisable to create a micro schedule of the change orders and then insert it into the updated baseline schedule to determine the full impact. A change proposal that was not on the critical path when submitted may now be.

Another way to manage late-stage change orders is to classify them as critical or non-critical. Naturally, critical change orders are necessary to Substantial Completion and non-critical changes can wait until after Substantial Completion. This can reduce the risk posed by the multiple change orders being executed at one time. Regardless of the method employed, each approved changed order should be plugged into the current schedule to determine its impact. This may involve an update to the current schedule with each change order added. The analysis can be used to build a case for constructive acceleration costs.

One final note about risk contingency planning: Risk is inherently increased as a project or task is crashed. The risk is not always linear and can sometimes change the dynamic of the task or project in undetectable ways. In Chapter 11 Acceleration and Schedule Compression, it was stressed that selecting a task to crash is often based on the cost to crash the task. There are factors that must be considered beyond simply cost. Some tasks pose higher risks than others when crashed. Crashing a more expensive task that may pose less of a risk should be an option for consideration.

Summary and Key Points

The primary responsibility of the project manager is to identify and manage risk. Each process or discipline in the project control spectrum focuses on reducing the negative occurrences that can sink a project. Despite the best efforts, risk can never fully be eliminated, so a plan to manage or minimize its impact must be developed and implemented if the event occurs. The primary take-away is that risk management is limited by the assumptions on which the calculations are based.

As with all parts of the project control process, risk must be recognized in the early planning stages so that a plan can be developed to address it in the most economical way possible. The concept of planning for risk helps compensate for the inevitability of some risks occurring. The risk identification process follows the Work Breakdown Structure as an outline to avoid omission. It forces a detailed analysis of what can go wrong and what is its impact, should it occur. Risk responses require a contingency funding plan to pay for the risk, should it occur.

Sophisticated computer analysis can use probability analysis to determine the best duration for a critical path task. This does come with limitations created by merge bias.

Experience has shown that using a structured process for dealing with the known and unknowns of the construction project minimizes delay, manages cost, and reduces personnel fatigue. In general, risk management adds value to the control process. Risk management is predicated on the project or company's tolerance for risk. How much confidence does the project need?

Chapter 14 Project Closeout will consider the final stages of project control that deal with the analysis of performance through the lessons learned process. In addition, all of the data from the project must be saved, archived, and made available as a guide to future performance so that mistakes are not repeated and achievements are.

Key points of this chapter are:

- There are three key steps to risk management: risk identification, risk assessment, and risk response.
- There are five ways in which risk can be managed: mitigate risk, avoid risk, share risk, transfer risk, and retain the risk.
- Any response to a risk needs a contingency plan and a source of funding to be successful.
- Durations of critical path tasks can use probability analysis software to aid in making decisions about time.
- Contract type will impact risk and project controls.
- Contract type can reduce or increase risk to the parties.
- Change orders issued near Substantial Completion can often change the critical path or the schedule and add risk.

Chapter 13 Risk Management: Questions for Review

1. All risk can be eliminated from a project through aggressive mitigation plans. True or False?

2. Risk can have a positive or negative effect on the project goals. True or False?

3. What is the purpose of the risk management process?
4. The first step in the risk management process is risk identification. True or False?
5. Probability is generally considered a more detrimental factor than impact. True or False?
6. What is the purpose of an impact scale?
7. The impact scale can then be used as the basis for a comparison that allows the project manager to assemble and rate the risk based on a numerical scale. This scale is called?
8. The Failure Mode and Effects Analysis (FMEA) introduces what additional factor to the severity matrix?
9. The process to reduce the likelihood that the event will occur and the impact of the event if it does is called?
10. PERT analysis of a task is used to identify the most probable duration of the task. True or False?

CHAPTER 14
PROJECT CLOSEOUT

The final step in the construction project life cycle is the project closeout. While project closeout occurs at the end of the project, planning for closeout starts in the initial pre-construction phase. The information and documentation needed for closeout are collected from the very early stages, beginning even before the execution of the work. The physical construction work is clearly the largest phase of the construction process; however, it is not the only phase. There is administrative and contractual paperwork that plays a large part in the transferring of the project from the construction team to the owner's team. Project closeout is not a complicated process, but it can be time-consuming.

Project closeout needs to be an orderly and organized process that transfers ownership and control of the project to the owner or end user. It should be discussed with the owner well in advance of closeout so that it educates the owner as to what to expect. A well-organized and complete closeout process adds to the owner's confidence in the project and can enhance the relationship for future projects. A poorly executed or lengthy, dragged out closeout process can reverse the good will generated through the construction process.

Our industry as a whole has struggled with project closeout, and much of it stems from not starting the process early enough. A closer look at the process may help.

Project Documentation

Project documentation is the history of the project. It covers a wide variety of paper and electronic documents that become the project record. As such, the collection of these documents starts at the very beginning with the estimate and bid documents themselves. Each piece of information that has contributed to the budget, means and methods, or schedule decisions must be saved for the record. Records generated during the pre-construction phase, construction phase, and closeout phase should also be retained.

The orderly and organized retention of records does not start when the project is 75 percent or 80 percent complete. It starts before the project starts. It starts with policies and procedures for retaining the information. This simple act will make the project closeout so much easier. Communicating to the team members where and how records are retained and stored, and where records can be found is part of the procedures.

Initiation of Project Closeout

Project closeout begins in the pre-construction phase of the project. It is initiated with a closeout meeting that lays out the contractual requirements with the subcontractors, vendors, and all participants in the process. While it may seem premature, it is quite the opposite. This early introduction to the project closeout process allows the team to develop a system for collecting the necessary documents and information during the execution rather than at the very end of the project. Many contractors require that their subcontractors include a project closeout line with commensurate dollar amount in their schedule of values. Some say that retainage is intended to ensure faithful execution of project closeout. This is incorrect.

A helpful technique to start the project closeout process is a careful study of Section 01 70 00—Execution and Closeout Requirements in the Division 1—General Requirements section of the technical specifications. Section 01 70 00 provides a detailed checklist of the items required for closeout compliance for each individual project. The section includes, but is not limited to:

- Closeout procedures
- Final cleaning
- Starting of systems
- Demonstration and instruction
- Protection of installed construction
- Punchlist and Certificate of Substantial Completion
- Project change order log
- Warranty items in process of correction
- Project record documents; final inspections, as-builts, etc.
- Closeout submittals
- Lien releases/Lien waivers
- Operations and maintenance data
- Preventative maintenance instructions
- Maintenance contracts
- Manual for equipment and systems
- Spare parts and maintenance products
- Testing and commissioning reports
- Product warranties and product bonds
- Surety bond releases
- Completed operations insurance policies
- Final site survey (if applicable)
- Extra stock materials (Attic Stock)

Once the project team is aware of the general requirements outlined in 01 70 00, the next step is to create a matrix of what is required for closeout in each of the technical specifications. This can often be found in Part 1—General of each technical specification section. It is recommended that the project team members review Part 2—Products and Part 3—Execution as well, to ensure nothing has been omitted.

The matrix should include a description of the closeout item, and also provide the submission date, date requested, date received and disposition of the closeout item, date forwarded to the owner/architect, and the name of the recipient. The project team member heading up the project closeout should always retain copies of all closeout submissions for archiving, as well as signatures acknowledging the receipt. While this may seem a bit drastic, more than one owner's team has misplaced an original document, which had to be replaced by the contractor or subcontractor because there was no proof of transfer. On the opposite side of the table, this prevents the dispute over what was received and what has not been received.

Project closeout procedures should be in place with much of the data collected before Substantial Completion. At Substantial Completion, many of the trades remobilize to more fruitful projects, leaving a diminished labor force. Occasionally, another team is selected to closeout the project, one with little knowledge or history of the project. Interest in the project is dissipated, personnel are fatigued, and frankly the project is in the rear view mirror. This is a recipe for disaster.

Objectives of the Closeout Process

The objective of the closeout process is to create an orderly and documented transition of the project, from possession by the construction team to possession by the owner or end-user.

To achieve a successful closeout, there needs to be acknowledgment that the closeout process means different things to different parties. To the owner it can often be a time of great anxiety and apprehension as the operation of the facility becomes their responsibility. Creating a smooth and orderly transition is the mark of a professional contractor, while a hasty exit can leave a lasting impression of abandonment. It is not uncommon for an owner to respond to the abandonment by holding onto the last vestiges of the contract—the contract retainage.

Architects with construction administration duties have often exercised the owner's contractual rights when the contractor has failed to focus on the completion of the punchlist and closeout procedures in a timely manner. These rights have included:

- The withholding of payment for failing to submit closeout documentation
- Refusing the release of retainage until all closeout documentation has been reviewed and accepted

- Assessing the prime contractor for the A/E firm's time and expenses for prematurely requesting Substantial Completion
- Extending the project closeout procedures ad infinitum
- Securing the services of another contractor to complete the punchlist and deducting the cost from the remaining balance

All of these actions can be counterproductive to a smooth transition, and may even result in a legal challenge. To keep all participants on target for a seamless project closeout, updates to the process should be incorporated in the regularly scheduled progress meeting minutes from the time the project is approximately 75 percent complete. In some circumstances, this may be later than ideal. There is no reason early subcontractors, for example, the formwork subcontractor for the foundation cannot have closeout documentation complete by the time the building is 75 percent complete.

Project closeout procedures can be separated into two main categories: contract closeout and administrative closeout. There is no specific preference as to which proceeds first. In fact, there is no reason that prevents the two from proceeding simultaneously.

Contract Closeout

Contract closeout involves the fulfillment of the obligations under the agreement that has governed the project. This includes both the prime contract with the owner and lower-tier contracts with subcontractors. It is also common to verify that the maintenance or service agreements that were part of the original contract are in order and scheduled. Contract closure is a simple but essential part of project closeout. It often involves a careful review of the governing contract to reconcile items such as allowances and unit prices that were part of the agreement. It should include a careful review of approved change orders to ensure contract sums are correct. This is also done as part of the administrative closeout. The duplication of this task in two different processes acts as a check and balance against error. Contract closeout may also include the matching of purchase orders to invoices to reconcile vendor payments. Any termination of vendors or subcontractors during the construction process should contain an explanation (reasons) of the termination and a reference to the supporting documentation in the event any actions arise from the termination. It should be noted that what seems like a sufficiently detailed explanation at the time of closeout may be woefully inadequate months or years later. A narrative is a method for ensuring that crucial details are not lost.

Administrative Closeout

The administrative closure procedure focuses on the collection, preparation, and archiving of administrative type documents. Administrative documents are best classified as the financial accounting summary: payment histories to ensure subs/vendors/suppliers have been paid, regular updating of the closeout matrix, the archiving of lower-tier documents, and the creation and distribution of lessons learned documentation.

Administrative closeout also includes the financial reconciliation of the project and the review of accounts, both internal and external, to verify they have been closed for the project. Approved change orders are tabulated to verify contract sums and payments made. Disputed amounts that have yet to be resolved are held in abeyance for later disposition. The final job cost report is produced and distributed for review.

Archiving Project Files

Historical project data is a priceless source of information for future projects. All project records—cost, schedule, record documentation—should be stored electronically according to the company record retention guidelines. As part of the archiving process, a detailed key to where specific files are stored must be developed. Whenever possible or appropriate, files should be categorized either chronologically or sequentially for ease of access at a later date. Archived files should contain at a minimum the following:

- Contract documents: plans and specification
- Contract modifications; change orders and CCDs
- Correspondence by recipient party and type
- Subcontractor contracts and modifications
- Purchase order logs for vendors and suppliers
- Transmittal logs
- Submittal logs; completed
- Applications for Payment
- RFI Log complete with responses
- Final cost accounting and payment record
- Baseline and regular updated schedules
- Meeting minutes; on-site and in-house
- Project closeout log
- Lessons learned report

It would be naive to not acknowledge that some projects end with unresolved issues that may require the involvement of a third party professional. For projects with unresolved claims, that may mean "going legal" and the documents must be made readily available. This should not, however, prevent some archiving from proceeding, the limits of which may be best discussed with counsel.

Claims and Disputes

As mentioned previously, not every project closes out "cleanly" or without *disputes*. These disputes can lead to a *claim*. A dispute as the name would imply is a disagreement between the parties. Given the nature of construction projects and the large number of subcontractors, vendors, and suppliers on any single project, it should come as no surprise that the parties may have different opinions. These differences, when not resolved in the normal course of events, can lead to a claim for an equitable adjustment to the contract.

The most common source of disputes are rejected change proposals. A prime contractor or subcontractor submits a change order request for work they see as outside the scope of their contract. The change proposal is summarily rejected by the owner through the architect (or engineer), and the prime contractor or subcontractor is directed to proceed with the work as base contract work. Remember, that the documents are subject to interpretation by the various parties using them. As such, most agreements provide that the architect (or engineer) is the interpreter of the design and renders clarifications on the design. On the face of it, this seems logical, given they designed the project and should know the intent of a line, symbol, or detail. However, it is not uncommon for the rejection and subsequent explanation not to satisfy the losing party. This rejection can generate a claim along with the notification paperwork required.

What happens next is crucial. More often than not there are discussions and negotiations that lead to a resolution within the framework of the contract. That is to say, there is no third party involvement. Claims are money and time bandits. The theft of money and time comes in the form of project management resources dedicated to resolving claims or worse, third party costs.

Sometimes claims are elevated to legal status when the parties want to pursue it beyond the project fence. Disputes and their claims that go legal can be expensive, mind-numbingly expensive, and should be avoided whenever possible. It is not uncommon for a project that has made a profit to see that profit dwindle in pursuit of a legal solution to a claim. Project managers should be tasked with the resolution of disputes at the project level before they become a claim. Should they mature into a claim, they should be directed to resolve them without third party involvement. Who is right and who is wrong is unimportant at this point. This is not a decision based on principles, but on finances. That having been said, there are times when disputes cannot be resolved and require outside expertise. This is the reason most contracts provide for mediation, arbitration, and litigation.

Lessons Learned Meeting

Lessons learned is the discussion and documentation of the experience gained on the project. Lessons are learned from working with or solving real-life problems that projects often give birth to. It is the review of the positive and negative situations and how the project team dealt with them throughout the life cycle of the project. It is intended to provide insight for future projects so that the positive experiences are replicated and improved and the negative experiences are not repeated.

Lessons learned meetings are typically conducted like any other meeting. An agenda is written and distributed well in advance for any additions by team members. Key stakeholders, such as the project team and the project manager, superintendent, senior executive managers, financial personnel, and, in many cases, stakeholders outside the immediate team are in attendance. Sufficient notice of the meeting should be provided to allow participants to collect their thoughts and create notes in advance. The meeting should be chaired by someone other than the project manager responsible for the execution of the project. This prevents a one-sided view of the project or glossing over of problem performance. For the lessons learned meeting to have value, the discussion must be frank and honest. Participants must be able to openly present their comments, both positive or negative, without fear of reprisals. Lesson learned sessions are not a finger-pointing opportunity, nor are they intended to be punitive.

Responsibility for problem performance is essential to developing useful recommendations for future projects of a similar nature. Lessons learned meeting are intended to focus on major issues, both bad and good. Participants should avoid personal attacks or "nit-picking" that might be interpreted as a personal attack.

One of the most commonly overlooked lessons to come from a completed project is the cost comparison analysis: the actual costs and how they were expended as compared to the estimated costs at the time of bid. The same is true about schedule. Both are often archived without the benefit of comparing the final costs to the estimated costs or final task duration to the estimated duration. Both offer key learning results that have the potential to improve future performance in either category. Sometimes this comparative analysis goes beyond the normal lessons learned meeting and may be best served by limiting it to the estimating and scheduling departments. This may also include the circumstances that contributed to the difference, if any. The project team must remember that not all lessons learned are tactical; some are simply arithmetical. It would be imprudent to continue to estimate a task at $5.00 per SF for labor if the estimator had data that supported an actual labor cost of $5.50 per SF. Use the data as a source of improving future performance.

Remember to recognize superior performance by members of the project team. Positive reinforcement can be extremely satisfying and rewarding to team members. Team members are generally proud to share positive public recognition by senior management when the project is successful. It can be an enormous motivator for future performance.

Lessons learned meetings can also be the best opportunity to discuss new techniques that were tried, products or processes used, or first-time subcontractors. These are critical bits of information that will help improve the second use of a product or subcontractor, or more importantly, prevent a repeat of a bad experience. Lessons learned sessions are a valuable closure mechanism for team members.

Lessons Learned Documentation

There are numerous formats for lessons learned documentation. Typically, each lesson can be committed to a separate piece of paper, or can be part of a single report for the project. The lesson learned can follow a narrative format, but must address the following:

- Describe in sufficient detail the issue that occurred.
- Summarize any triggers or thresholds for recognizing that the issue was occurring.
- Outline in detail the impact (cost and/or schedule) of the issue.
- Outline in detail the root cause of the issue.
- Provide references or sources of information for diagnosing the issue or applying the corrective actions (if a negative issue).
- Identify the corrective actions taken and the results needed to remedy the problem.
- Provide any actions for exploiting the issue (if a positive issue or opportunity).

Lessons learned and comments regarding project assessments should be documented in detail and openly discussed with the intent of eliminating future occurrences of avoidable issues.

Project Closeout Report

The project closeout report documents the completion of the closeout process. The report provides a historical summary of the project: baselines, achievements, and most important lessons learned as a result of its execution. The report should identify variances from the baseline costs and schedule metrics set at the planning phase of the project. The project closeout report is intended as a frank, concise evaluation of the project and the performance of its team.

Team members are encouraged to provide input to the project manager who generally authors the report. People who perform different functions during the process may have very different perspectives of the successes and failures of the project and, as a result, may provide helpful insight for future projects.

The closeout report is written after the project is fully completed and all closeout tasks have been finalized, yet while memories of the project are still fresh in the mind of team members. Sixty to ninety days after closeout is a typical window for creating the report.

Summary and Key Points

Of equal importance to the planning phase of the project is the closeout of the project. To prevent critical information learned during execution from being forgotten, the closeout process begins almost immediately with the start of the work. Project closeout cannot be done efficiently without a structured company policy or procedure for retaining the project documentation. Setting specific criteria for monitoring and reporting closeout early on helps create value in the process and eliminates the crisis management that occurs in the absence of a protocol. Establishing a plan early helps prevent "holes" or the inevitable omissions that occur near the end of each project.

Project closeout can be classified into two categories: contract and administrative, each with its own field of focus. Overlapping duties in each can often provide a check and balance to the process, thereby ensuring a complete transition. The final documentation is archived for future use.

The closeout process is aimed at creating a smooth transition between the contractor and the owner during the handover. This handover must be orderly and professional and must demonstrate the contractor's commitment right to the end.

Projects generate disputes and some disputes can end in claims against the contract. Whenever possible, both the dispute and the claim should be addressed and resolved between the parties involved before turning to outside help such as attorneys, mediators, arbitrators or the court system. Matters that go legal can be huge wastes of time and money.

Ideally, each project imparts some practical wisdom to the participants. It is the by-product of experiencing and solving real-life problems or creating opportunities that are part of every project. Documenting and sharing this wisdom to help improve future projects are part of the lessons learned process and crucial to the growth of any project manager or company he or she serves.

Key points of this chapter are:

- An organized project closeout is extremely important to maintaining the professionalism of the project.
- The retaining and organizing of project documentation happen as a result of policies and procedures by the contractor.
- Project closeout starts very early in the project.
- There are two phases of closeout: administrative and contract.
- Claims and disputes waste time and money and should be resolved as early as possible.
- Project records generate information that should be shared in a lessons learned meeting or report.

Chapter 14 Project Closeout: Questions for Review

1. What is the objective of the project closeout process?
2. Project closeout begins at the end of the project. True or False?
3. Project closeout procedures can be separated into two main categories. Define each.
4. Lessons learned meetings are used to avoid repeating the same mistakes and to improve performance on the project. True or False?
5. Explain how job cost reports are part of the project closeout and lessons learned meeting.
6. Name five items that one would expect to see in lessons learned documentation.
7. Why is it essential for the report to contain viewpoints other than those of the project manager?
8. What is the purpose of archiving files?
9. Explain where a project team member would find what is required for closeout.
10. The estimator(s) should be part of the project closeout process. True or False?

Case Study

Maybe one of the best ways to reinforce learning is to demonstrate it in a real-life application. This chapter is an attempt to do just that. The background has been laid out for a particular scenario and much of the decision making has be done in the pages that follow. While it is not an actual situation from past experience, it is very similar and naturally the names have been changed to protect the innocent, and those not so innocent!

Owner/Developer Background

Hometown Development, Inc. (HDI) is a large regional development company in the greater Boston area. They have been in business for over 40 years with an excellent reputation among its clientele. They are privately held. In the industry, the reputation is as a company that pays its bills but a bit of a "hard ball" player. That is to say that they play by the rules and do not forgive contractor-caused errors or delays easily. Projects are multi-year projects. They work with the best design teams in the area, and on this project the design team is Adams Munroe and Associates (AMA). AMA, including its subconsultants, has a reputation of producing comprehensive, well-coordinated, detailed project documents. They typically have few RFIs in the bid phase. HDI is a strong believer in fully developed project documents to avoid change orders as a result of ambiguities or disputes.

HDI specializes in apartment buildings, ranging from economy-based units to luxury units, based on the demographics. They operate in a 100-mile radius of the city and do extensive research before going forward with land acquisition or design. Their designs use top-end materials to minimize maintenance as they retain and manage many of their projects when complete. Their projects are highly desirable among those looking to lease and always are at the upper end of any market they are in.

Contractors that bid their work are prequalified, including financial statements, references, bonding, etc. Bidders are invited from a select list. Projects have a maximum of four bidders with excellent qualifications and reputations. Bidders are almost exclusively local or regional companies.

Project Background

The project is Whistling Pines Residences and is a multiphase project structured over three years. It consists of eight buildings of residences, containing four floors each, with ten apartments per floor. In addition there will be a 10,000 SF community structure with gym and banquet facilities. This is a total of 320 residential units ranging in gross area from 1,000 SF, 1,350 SF, and 1,650 SF based on a normal mix of 1-, 2-, or 3-bedroom units. Each building is the same layout per floor, with mirrored configurations. The gross area per floor is 16,000 SF with a total of 64,000 SF per building. Phase 1 is three residential structures, Phase 2 is three residential structures, and Phase 3 is two residential structures and the community center. The sitework will progress with each phase.

Buildings are wood-framed and sheathed with cement fiber siding and synthetic trims. The roof is truss framed with plywood sheathing and an asphalt shingled roof. The buildings are approximately 200' x 80' with concrete masonry unit (CMU) stair towers at each end of the structure, concrete-filled metal pan stairs. There are two elevators per building with CMU hoistways. Composite floor joists were used with a mix of conventional lumber and laminated veneer lumber (LVL), sheathing, etc. With the exception of the CMU stair towers and hoistways, the structure is a wood-framed building.

Each phase of the project will be bid and awarded separately. Phase 2 will follow Phase 1. Award will be based on price and best value. There may be delays between phases allowing the structure to be fully leased, and subsequent phases may be postponed should the rental market soften. The hard cost budget for Phase 1 was estimated to be approximately $51.9 million if performing the work with an open shop labor force. The Total Allocated Budget with Management Reserve is just shy of $52.5 million. HDI estimated an 18-month schedule for Phase 1 to target the rise of the rental market. HDI and AMA have permitted the project and are ready to break ground immediately upon award, which will be on or about April 10, 2023. Contractors were provided a 30-day bid period beginning January 16, 2023, with bids for Phase 1 due on February 16, 2023. Award will be on or about March 10, 2023. The lender/investors for Phase 1 funding are in place.

Whistling Pines is at the upper end of the project size (all phases combined) for HDI, but well within their management control. AMA and its subconsultants have sufficient staff to administer the project. This Case Study will concern itself with Phase 1 only.

General Contractor Background

Atlantic Construction, Inc. (ACI) is a privately held open shop company that has been in business for over 60 years. The principals of the firm are two brothers, Bob and Tom, both in their sixties and are third-generation management of the firm. Fourth-generation family members are being groomed for takeover when the two principals retire, they are Mike the estimator, Bob Jr. the project manager, and Lisa is the CFO. The company is bonded and has a single project limit of $55 million with an aggregate of $90 million. Senior management have been with the firm for a minimum of 11 years and some as long as 22 years. ACI is a financially stable company, but conservative in their business strategies. The principals have a low tolerance for risk. They do a mix of private commercial and public (taxpayer-funded) work, with a gross volume of $110–$120 million per year.

ACI self-performs most carpentry tasks: rough and finish. They subcontract the majority of large subtrades. They have completed projects similar to this before, three with HDI, yet this would be the largest. They have done several projects with AMA and have a professional relationship with both firms.

The project superintendent, Mark, is another family member with 30 years of experience. He has been with ACI since college graduation as a civil engineer and is a key decision maker in subcontractor selection.

They bid Phase 1 and after a couple rounds of negotiations were awarded the project for $51.2 million with a schedule of 510 calendar days (approx. 17 months) from Notice to Proceed to Substantial Completion. Both price and schedule are considered aggressive by ACI senior management, but achievable based on past projects. Their major subcontractors are on board and they are comfortable with their estimate, having reviewed it during negotiations. The estimate was a prepared by Mike, but Bob Jr., Lisa, Mark, and Woody contributed.

During pre-construction planning, the ACI team risk analysis revealed that, in addition to buying the subcontractors, vendors, and materials suppliers, thus maintaining control over the subcontractors' schedule, the key to both time and cost will be the wood framing of the structures. This is a critical path task and the duration must be maintained on Building #1 and improved in Buildings #2 and #3 by mastering any learning curve on Building #1. Also material pricing for lumber and sheathing have been secured at a distributor level to stabilize prices to a maximum increase of 6 percent between Buildings #1 and #2 and 3 percent between Buildings #2 and #3. The wood framing scope is seen as the largest schedule risk to the project. The price of lumber is the largest cost risk to the project.

The framing crew has been with ACI for multiple years and are led by a time-tested and competent foreman, Woody (yes. . . I said Woody). Woody has worked for 11 years with ACI. He has a vested 401-k retirement package and profit-sharing program with ACI. He is by definition a "company man" and has the ear of senior management. Woody is a "walking" foreman vs. a "working" foreman. He collaborated with Mike to define the various risks and durations of each step of the framing and also solicited additional carpenters to interview and add to the framing crew.

Factors Affecting Construction

The economic conditions of the area for construction are stable but not robust. Prices favor the owners (HDI). Material prices are in flux due to supply chain economics. Labor is available and sufficient for most trades.

Lumber and sheathing prices are a concern. The current administration in Washington is weak on business-friendly policies and Massachusetts is in lock step with Washington. There is little hope that the administration will solve the economic problems during Phase #1 that has allowed the supply chain dilemma to drag on.

The site has been cleared of debris from previous structures, above and below grade and treed areas have been cleared and stripped of top soil under a separate contract. There are no anticipated surprises below grade and testing and geotechnical reports support this premise. The site is predominately glacial till and of structural grade quality soil. A graded building pad has been provided for the start of Building #1.

April in the greater Boston area is prone to approximately 15 days of rain. Time extensions will be granted for rain days in excess of: 15 days in April, 10 days in May, 8 days in June.

The site subcontractor for the project is the largest in the area and has sufficient equipment to maintain the progress of Phase 1 as defined by the ACI's CPM schedule.

Plan for Phase 1

During the planning phase it was decided that the formwork for the foundations would be split between two companies. Building #1 and #3 would be by one company and Building #2 would be done by a smaller form contractor. Phase #1 would start with the excavation of the Building #1, proceed to #2 and then #3. The larger formwork company will start with Building #1. Part of the plan was that each company would provide labor to the other should there be delays due to weather or production issues.

A backup plan to assure no delays with the foundation. The foundations are full basements (50 percent) and crawls spaces (50 percent).

As a means of minimizing delay, language was included in the subcontracts that if a day was lost to weather during the week, the crew would work Saturday to offset the delay. There would be no premium wages for this remedy. This policy was also true for the framing crews.

As noted previously, the vendor for the lumber was secured at a distributor level. While there was considerable concern that this would foster bad feeling among local lumber yards, none of the lumber yards contacted would commit to supplying the project needs in full. This was also a condition of the negotiations between HDI and ACI. The same was done with the asphalt roof shingles, siding and exterior trims, windows, and exterior doors.

It was also decided that the truss fabricator would start immediately on the shop drawings and they would be built using hold-to dimensions after approval of engineering calculations. The truss fabricator has demonstrated their ability to meet this type of schedule, having done business with ACI and HDI in the past.

The framing crew would not set the roof trusses or sheath the roof. This was subcontracted out, again only for Building #1 to start. The plan was to see how well the framing crew met the goals set below the roof. There was no plan in the budget (estimate) for the use of overtime and premium wages. A mill was set up for cutting and assembling headers, corners, studs, etc. and materials were staged adjacent to the building and handling will be done with a telescopic all-terrain forklift (Highlander).

Schedule for Building #1

After a detailed review of the estimate, the team including field personnel met to discuss the schedule for the wood framing of each building. The team set goals for framing and sheathing each floor and setting the trusses and sheathing the roof. The goals were based on days. The decision was made that they would start with a 12-member framing crew (Crew A) and a second 9-member crew (Crew B) to do the pick up on each floor. Crew A is composed of four lead carpenters and eight carpenters/laborers, Crew B three lead carpenters and six carpenter/laborers. Woody would be additionally overseeing all crew members. Goals for the crew to frame and sheath are identified in Figure CS.1 Framing Schedule for Building #1.

The team decided that they would review progress and cost weekly to compare it to the plan and make any adjustments as needed.

Ref. no.	Task description	Crew	Planned duration (days)
1	Frame and sheath first floor deck over basement and crawl space areas including support members below	Crew A	5
2	Frame and sheath exterior walls and interior load bearing partitions on first floor	Crew A	4
3	Frame and sheath second floor deck over full footprint	Crew A	5
4	Frame and sheath exterior walls and interior load bearing partitions on second floor	Crew A	4
5	Frame and sheath third floor deck over full footprint	Crew A	5
6	Frame and sheath exterior walls and interior load bearing partitions on third floor	Crew A	4
7	Frame and sheath fourth floor deck over full footprint	Crew A	5
8	Frame and sheath exterior walls and interior load bearing partitions on fourth floor	Crew A	4
9	Truss, brace, and sheath roof frame over footprint	Subcontractor	5
10	Frame interior NLB partitions and strap ceiling at first floor	Crew B	4
11	Frame interior NLB partitions and strap ceiling at second floor	Crew B	4
12	Frame interior NLB partitions and strap ceiling at third floor	Crew B	4
13	Frame interior NLB partitions and strap ceiling at fourth floor	Crew B	4
Total Duration in Workdays			57

Figure CS.1 Framing Schedule for Building #1

Budget for Building #1

Wages were fixed for the framing crew members for Phase 1. The tracking costs were based on wages, fixed taxes and insurances, worker compensation, general liability, etc. All costs with the exception of indirect overhead and profit are in the hourly rate. The crew costs for Crews A and B are identified in Figure CS.2 Framing Budget for Building #1.

Ref. no.	Task description	Assigned crew	Planned labor cost
1	Frame and sheath first floor deck over basement and crawl space areas including support members below	Crew A	$ 25,440.00
2	Frame and sheath exterior walls and interior load-bearing partitions on first floor	Crew A	$ 20,352.00
3	Frame and sheath second floor deck over full footprint	Crew A	$ 25,440.00
4	Frame and sheath exterior walls and interior load-bearing partitions on second floor	Crew A	$ 20,352.00
5	Frame and sheath third floor deck over full footprint	Crew A	$ 25,440.00
6	Frame and sheath exterior walls and interior load-bearing partitions on third floor	Crew A	$ 20,352.00
7	Frame and sheath fourth floor deck over full footprint	Crew A	$ 25,440.00
8	Frame and sheath exterior walls and interior load-bearing partitions on fourth floor	Crew A	$ 20,352.00
9	Frame interior NLB partitions and strap ceiling at first floor	Crew B	$ 15,264.00
10	Frame interior NLB partitions and strap ceiling at second floor	Crew B	$ 15,264.00
11	Frame interior NLB partitions and strap ceiling at third floor	Crew B	$ 15,264.00
12	Frame interior NLB partitions and strap ceiling at fourth floor	Crew B	$ 15,264.00

Note: Truss erection and roof sheathing have been omitted.

Figure CS.2 Framing Budget for Building #1

Figure CS.3 Planned Values for Cost and Schedule is the combination of time and cost for the as-planned values for each scope.

Progress and Cost Report for Building #1

Figures CS.4–CS.8 are the progress reports generated at the end of each week comparing the as-planned progress to the actual progress. Also included are the cost reports for the same time period. Both reflect labor only, as much of the material costs have been posted to the cost account and the materials are staged adjacent to Building #1.

Ref. No.	Task description	Crew	Daily rate	Duration	Cost
1	Frame and sheath first floor deck over basement and crawl space areas including support members below	Crew A	$5,088	5	$25,440
2	Frame and sheath exterior walls and interior load-bearing partitions on first floor	Crew A	$5,088	4	$20,352
3	Frame and sheath second floor deck over full footprint	Crew A	$5,088	5	$25,440
4	Frame and sheath exterior walls and interior load-bearing partitions on second floor	Crew A	$5,088	4	$20,352
5	Frame and sheath third floor deck over full footprint	Crew A	$5,088	5	$25,440
6	Frame and sheath exterior walls and interior load-bearing partitions on third floor	Crew A	$5,088	4	$20,352
7	Frame and sheath fourth floor deck over full footprint	Crew A	$5,088	5	$25,440
8	Frame and sheath exterior walls and interior load-bearing partitions on fourth floor	Crew A	$5,088	4	$20,352
9	Frame interior NLB partitions and strap ceiling at first floor	Crew B	$3,816	4	$15,264
10	Frame interior NLB partitions and strap ceiling at second floor	Crew B	$3,816	4	$15,264
11	Frame interior NLB partitions and strap ceiling at third floor	Crew B	$3,816	4	$15,264
12	Frame interior NLB partitions and strap ceiling at fourth floor	Crew B	$3,816	4	$15,264
Total Costs and Duration				52	$244,224

Note: Truss erection and roof sheathing have been omitted.

Figure CS.3 Planned Values for Cost and Schedule

		Progress		Schedule		Cost	
Ref. no.	Task description	Planned value complete	Actual value complete	Planned duration	Actual duration	Planned cost	Actual cost
1	Frame and sheath first floor deck over basement and crawl space areas including support members below	16,000 SF	15,480 SF	5 days	5 days	$25,440.00	$25,440.00

Figure CS.4 Cost and Schedule Report, Week #1

		Progress		Schedule		Cost	
Ref. no.	**Task description**	**Planned value complete**	**Actual value complete**	**Planned duration**	**Actual duration**	**Planned cost**	**Actual cost**
1	Frame and sheath first floor remaining deck from Week #1	0 SF	520 SF	0 days	0 days*	$ -	$ 1,526.00
2	Frame and sheath exterior walls and interior load bearing partitions on first floor	960 LF	960 LF	4 days	4 days	$20,352.00	$20,352.00

Note: * The remaining deck from Week #1 was completed by Crew B as a fill-in.

Figure CS.5 Cost and Schedule Report, Week #2

		Progress		Schedule		Cost	
Ref. no.	**Task description**	**Planned value complete**	**Actual value complete**	**Planned duration**	**Actual duration**	**Planned cost**	**Actual cost**
3	Frame and sheath second floor deck over full footprint	16,000 SF	15,600 SF	5 days	5 days	$25,440.00	$25,440.00

Figure CS.6 Cost and Schedule Report, Week #3

		Progress		Schedule		Cost	
Ref. no.	**Task description**	**Planned value complete**	**Actual value complete**	**Planned dduration**	**Actual duration**	**Planned cost**	**Actual cost**
3	Frame and sheath second floor deck of remaining area from Week #3	0 SF	400 SF	0-days	0.25-days*	$ -	$ 954.00
4	Frame and sheath exterior walls and load-bearing partitions on second floor	960 LF	960 LF	4-days	4-days	$ 20,352.00	$ 20,352.00
9	Frame interior NLB partitions and strap ceiling at first floor	16,000 SF	16,000 SF	4-days	3.6-days	$ 15,264.00	$ 13,780.00

Note: * The remaining deck from Week #3 was completed by Crew B before moving on to interior partitions on Floor 1.

Figure CS.7 Cost and Schedule Report, Week #4

		Progress		Schedule		Cost	
Ref. no.	**Task description**	**Planned value complete**	**Actual value complete**	**Planned duration**	**Actual Duration**	**Planned cost**	**Actual cost**
5	Frame and sheath third floor deck over full footprint	16000 SF	16000SF	5 days	4 days*	$ 25,440.00	$ 29,256.00
10	Frame interior NLB partitions and strap ceiling at second floor	16000 SF	16000 SF	4 days	4 days	$ 15,264.00	$ 15,264.00

Note: * The third floor deck was framed in 4 days by adding Crew B for 1 day. The third floor deck and the second floor interior partitions were staggered by 1 day, which prevented interference.

Figure CS.8 Cost and Schedule Report, Week #5

Consider the labor costs and physical progress of the work and review the efficiency for both cost and schedule (CPI and SPI).

In analyzing the work, what are some key indicators that Bob Jr. the project manager should be looking for? Identify the key indicators and prepare a brief outline of what you would report to ACI and HDI senior management as far as progress and forecasting future performance are concerned.

Case Study: Questions for Group Discussion

The following are some questions for discussion in a group or classroom setting. Discussion and responses should be based only on the information provided, or what could be reasonably concluded from it. Assume the role of project manager for ACI.

1. What would you consider to be ACI's tolerance for risk? High, medium, low? Explain your conclusion.
2. How would you define ACI financially? What are the indictors? How will this impact the work?
3. Discuss the senior management of ACI. What can be concluded from the information provided? What further questions should be asked?
4. How do you think the project will be impacted by the prior relationship between the parties: HDI, AMA, and ACI? Can that be relied on?
5. What are the SPI and CPI for each cost report? Is there a trend that can be forecast? How reliable is the trend?

6. What was the point in fixing the framing crew's wages for Phase #1?
7. Discuss your agreement or disagreement with the lack of overtime in the budget. If you feel overtime should have been included, on what basis would it be estimated?
8. What plan would you (as the PM) put in place to handle the natural attrition that will occur with the framing labor force?
9. Discuss the pros and cons of self-performing the framing.
10. Discuss the pros and cons of subcontracting the setting of the trusses.
11. Comment on HDI's policy of preferring "fully developed project documents to avoid change orders as a result of ambiguities or disputes" as it applies to the framing scope.
12. Discuss the advantages and disadvantages of having two formwork subcontractors and how the work has been shared.
13. Identify and discuss any risk-response plans that have been made. Opine as to the benefit they provide or how they handle the risk.
14. What could be reliably reported to HDI and AMA at a meeting at the beginning of week #4?
15. What adjustments, if any, would you make to the primary framing crew?
16. Identify and discuss schedule or cost concerns you would have as the project manager. Opine as to potential remedies for those concerns.
17. The framing crew failed to meet the production for Week #1. What would you check that may have caused the failure?
18. Discuss how the cost and schedule data can be used to refine Building #2 schedule. How could it be used should ACI decide to bid Phase #2 and/or #3?

APPENDIX

QUESTIONS FOR REVIEW AND ANSWERS

Chapter 1 The Basics

1. Name the two key principles described in Gantt's 1919 book *Organizing for Work.*

1. (1) Measure activities by the amount of time needed to complete them (task duration); and (2) the space on the chart can be used to represent the quantity of the task that should have been done in that time (daily output).

2. Gantt was one of the first to recognize that in order for a team of workers to produce efficiently and to maintainable standards, what was required?

2. The team needed an intelligent leader who could solve or preempt problems.

3. Fayol's book *General and Industrial Management*, published in 1916, outlined a flexible theory of management that could be applied to most industries. What was the theme of the book?

3. The book stressed the importance of planning as a way to improve efficiency.

4. Gantt is recognized for what visual tool which measures actual progress against planned progress?

4. The Gantt Chart.

5. Name the six management functions that Fayol is credited with that are still the basis of project management today.

5. (1) Forecasting; (2) Planning; (3) Organizing; (4) Commanding; (5) Coordinating; and (6) Monitoring.

6. Identify the five steps in the traditional approach to project management.

6. In the traditional approach there are five distinct steps: (1) Initiation; (2) Planning; (3) Execution; (4) Monitoring and Controlling; and (5) Completion.

Project Control: Integrating Cost and Schedule in Construction, Second Edition. Wayne J. Del Pico.

7. Name the five approaches to managing projects.

7. The five approaches to project management are: (1) traditional; (2) critical chain project management; (3) extreme project management; (4) agile project management; and (5) event chain methodology.

8. What is the most tangible benefit of the pre-construction Planning Process?

8. Pre-construction planning forces detailed thinking about the project and the assumptions made during bid time.

9. Define the components of the Monitoring and Controlling process.

9. The Monitoring and Controlling process consists of a series of steps to observe the executing process. It is the establishment of a guidance system or a set of metrics by which actual performance can be measured and compared to planned performance.

10. Explain the role of the contract documents in the control process.

10. The contract documents form the basis of the performance measurement baseline. It is the basis of the costs and schedule model.

Chapter 2 Introduction to Project Control

1. Project control is the function of integrating cost and schedule data to establish a baseline or guidance system for monitoring, measuring, and controlling performance. True or False?

1. True. It is a two-front system for determining progress.

2. Control systems are established for what six factors in the management process?

2. Control systems are established for: (1) schedule; (2) cost; (3) contract modifications; (4) risk; (5) quality; and (6) resources.

3. The gathering of information that will be used during the analysis portion of the control process is called?

3. Project performance measurement. It is the regular and systematic collection of data to be used to determine progress and cost performance.

4. What is the name of the time and cost parameters that the project team sets as the metrics to measure performance?

4. The performance measurement baseline is the time and cost parameters that the project team sets as the metrics to measure and compare performance against.

5. Deviations from planned performance are called?

5. Variances. Variances are a key indicator of performance.

6. What is the name of the decomposition process that breaks the project down into its element components of work?

6. Work Breakdown Structure (WBS). It is the breakdown and organization into logical groups of tasks that can be tracked as one work package.

7. Identify the acronym ACWP and provide a definition.

7. ACWP is the Actual Cost of Work Performed. ACWP is the actual cost incurred in the performance of the work, or Earned Value for a given time period.

8. Identify the acronym BCWS and provide a definition.

8. BCWS is the Budgeted Cost of Work Scheduled. BCWS is the value of the work planned or scheduled to be accomplished within a specific time period as illustrated. It is the anticipated progress of the task.

9. Identify the acronym BCWP and provide a definition.

9. BCWP is the Budgeted Cost of Work Performed. BCWP is the value of the work completed measured in terms of the planned value of the work. It is the Earned Value for a task.

10. Define the formula for Schedule Performance Index (SPI) and explain what it is a measure of.

10. Schedule Performance Index (SPI) = BCWP ÷ BCWS. SPI is the budgeted cost of work performed (BCWP) divided by the budgeted cost of work scheduled (BCWS). It is a ratio of the earned value to the planned value. Values greater than 1.0 are considered favorable, values less than 1.0 are unfavorable.

Chapter 3 Pre-Construction Planning

1. Successful construction companies have policies and procedures for managing a project. Policies and procedures are designed to control an action in order to provide a predictable outcome. True or False?

1. True. Those actions are intended to further the performance of the work.

2. The pre-construction phase is intended to evaluate the constructability of the project and the documents that will be used as the guidance system for that process. True or False?

2. True. It forces detailed thinking about the project.

3. Define the responsibilities of the project manager.

3. The project manager is the lead on the project team. He or she is the decision-maker in the feedback cycle and is responsible for overall project performance; success or failure. They report to senior management, and are responsible for all team members receiving regular updates on project status. In many companies they are also the project scheduler and the project controls analyst. The PM works with the estimator and other team members to arrive at the Detailed Cost Accounts that will be the basis of the project cost

control. The PM assembles the periodic billing, reviews it with the architect or owner's rep and sees to it that it is approved and paid. The PM also approves or rejects payments to subs, vendors, and suppliers. Any value engineering initiatives, while not always initiated by the PM, are directed by the PM. The project manager chairs the meetings in the office and on-site. With the contributions of team members, the PM sets the meeting agenda to ensure that pertinent issues are discussed and resolved in a timely manner.

4. Define the term "means and methods".

4. Means and methods is the "detail behind how the project is to be built." It is the action that results in the execution of the plan that will fulfill the contract obligations. It is based on the means and method that was the basis of the estimate.

5. Many of the costs associated with a construction project are time-sensitive. Define what is meant by time-sensitive in this context and provide some examples.

5. Time-sensitive means that the longer a cost is exposed on a project, the greater the amount. Superintendent's salary is an example of a time-sensitive costs. The longer the superintendent is on the project site, the more he or she costs.

6. Explain the difference between the plan and the schedule.

6. The plan is the roadmap as to how the work will be executed. It is the selected means and methods to perform the work. The schedule is the plan set to time.

7. What is the baseline for time management and control on a project? Explain.

7. The initial schedule is the baseline for time management and control. It sets the standard by which schedule performance is measured.

8. The initial schedule that is published for use as a management tool is called?

8. The initial schedule is called the as-planned, baseline, or target schedule. It is the performance baseline for the time portion of project control.

9. The project estimate is broken down into costs for measuring fiscal performance. This breakdown process is called?

9. The Cost Breakdown Structure or the CBS.

10. Name three key components of a good communication plan.

10. (1) The information needs required by the group; (2) source of the information; and (3) the frequency of the distribution of the information.

Chapter 4 The Schedule

1. The construction schedule measures work as a function of time. True or False?

1. True. It is the work of the estimate over time.

2. The incremental parts or work of the schedule are called?

2. Tasks or activities.

3. What are required to measure incremental progress toward completion?

3. Milestones, or interim goals along the path to Substantial Completion.

4. A velocity diagram is used for scheduling what type of work?

4. Velocity diagrams are a type of schedule used for work that is linear in nature without clear segments of work that can be analyzed. Construction of a highway is an example of a linear type of project.

5. What type of scheduling assumes a high degree of variability in task durations?

5. Program Evaluation and Review Technique or PERT.

6. Define the basic concept of CPM scheduling. Cite the relationships between tasks and why it is pertinent in construction.

6. The theory behind the CPM schedule is that there are a series of sequential tasks that are linked from the start to the end of the project. If any one of these tasks is delayed, it will delay the finish date of the project as a whole. It is based on the fact that these tasks, called critical tasks are separate but inextricably linked by their interdependency.

7. Define the term "float." Explain how it can benefit the project.

7. Float is measured in workdays. It is the difference between the time required to perform the task and time available to perform the task.

8. Tasks can be classified into three types. What are the three types?

8. Tasks can be classified into: (1) administrative tasks; (2) procurement tasks; and (3) production tasks.

9. Task durations are measured in workdays, overall schedules are measured in calendar days. Explain the difference.

9. Workdays are typically Monday to Friday, exclusive of holidays or other non-working days. Workdays can be Monday to Sunday if the project runs seven days a week. Calendar days are Monday to Sunday including holidays and non-working days.

10. Explain the difference between the Daily Production Rate method and the Labor-Hour Productivity method for calculating durations.

10. The Daily Production Rate method is based on the productivity of a specific crew and the assumption that the productivity of this crew will remain constant over the life of the task, provided that the conditions under which the work is performed do not change. The Labor-Hour Productivity focuses less on the crew size, than the time it takes to install a single unit of the task. It is based on the fact there is a direct correlation between the cost and the labor-hours to perform the task.

Chapter 5 The Budget

1. Define the term "cost" and explain how it is affected by its frame of reference.

1. Cost is defined as the price paid to acquire, produce, accomplish or maintain something. It varies based on its frame of reference. For the owner, who will pay the general contractor, cost includes the contractor's profit. For the general contractor who will perform the work, cost is everything but the profit.

2. Specifications are qualitative in nature and set the acceptable standards that the quality of the project will be measured against. True or False?

2. True. Used with the plans, they create a quantitative and qualitative basis for the project.

3. Which type of estimating is the methodology by which bid documents are broken down into their incremental components for pricing?

3. Unit price estimating is the decomposition of the project into tasks that can be estimated.

4. The compilation of the estimated costs for a project becomes the budget for the project. True or False?

4. True, the budget is derived from the estimate.

5. Explain what CSI MasterFormat is and how it is used in the construction industry.

5. CSI MasterFormat is an organization structure that groups similar types of tasks in sections. The technical specifications for CSI MasterFormat and consequently the estimate are organized by its divisions and sections.

6. What type of information is contained in Part 1 General section of a technical specification section?

6. Part 1, the General section of the specifications, provides a summary of the work included within that particular section. It ties the technical section to the General Conditions and Supplementary General Conditions of the Contract, an essential feature in maintaining continuity between the general contractor and subcontractors. Part 1 identifies the applicable agencies or organizations by which quality assurance will be measured. It defines the scope of work that will be governed by this technical section, including, but not limited to items to be furnished by this section only, or furnished by others and installed under this section. It also identifies other technical sections that have potential coordination requirements with this section, and defines the required submittals or shop drawings for the scope of work described in this section. Part 1 also establishes critical procedures for the care, handling, and protection of work within this section, including such ambient conditions as temperature and humidity.

7. By what methods are products contained in the technical specifications specified?

7. Products can be specified by: (1) name or proprietary specifying; (2) performance specifying; (3) descriptive specifying; and (4) specification compliance number.

8. What is a cost element?

8. A cost element is the lowest level of the cost breakdown structure (CBS). It is material, labor, equipment, labor-hours, and equipment hours for a task.

9. What is the purpose of the project job costs report?

9. The job costs report provides the project manager/team with a regular summary of the costs incurred on a project in performance of the work.

10. Additions or deletions from the original budget by approved change order are reflected in the updated budget called?

10. The current budget. The current budget will change over the life of the project.

Chapter 6 Integrating the Schedule and the Budget

1. What is a Schedule of Values and what does it represent?

1. The Schedule of Values (SOV) is the decomposition of the contract into tasks or groups of tasks that have a dollar value assigned to them. The total of the SOV must equal the contract price including the fee.

2. The full value of a line item in the Schedule of Values represents the value of the scope of work at 100 percent complete. True or False?

2. True. It is the value of the task when complete.

3. The practice of placing a greater percentage of profit, or inflating the value of tasks, occurring early in the project is called?

3. Front loading the schedule of values. It is intended to offset the retainage.

4. What are the five values that must remain constant in order to compare actual to planned value?

4. (1) All tasks are being performed by professional tradespersons; (2) the tradespersons have the appropriate materials, tools, equipment, and training to perform the task; (3) the crew assigned to perform the task is the optimal crew. All crew members are contributing to the production; (4) work is being performed continuously without any interruptions or delays; and (5) the work-day remains constant 8 hours per day.

5. The Schedule of Values is used in determining the value of the work that has been completed in a specific period. True or False?

5. True. It serves as the basis to which the percentage of completion is applied.

6. The scheduled value of the task distributed over the duration of the task is called?

6. The planned value (PV) of the task.

7. The dollar value of the work scheduled to be accomplished in a specific time period is called?

7. The Budgeted Cost of Work Scheduled (BCWS). It is also known as the Planned Value.

8. The BCWP is the actual dollar amount earned based on what was scheduled to be earned. True or False?

8. True. It is also called the Earned Value or EV.

9. Combining the scheduled value of the work in dollars from the Schedule of Values with the anticipated progress predicted in the schedule, the project manager can forecast earnings from this task at any point along the schedule. True or False?

9. True. It can be used to predict cash flow.

10. Average daily production can be used to forecast future production if conditions remain unchanged. True or False?

10. True. Past performance is good indicator of future performance under these conditions.

Chapter 7 Calculating and Analyzing Progress

1. There are generally six recognized methods for calculating progress on a construction task. Identify all six.

1. (1) Units Completed; (2) Incremental Milestones; (3) Start/Finish; (4) Cost Ratio; (5) Experience/Opinion; and (6) Weighted or Equivalent Units.

2. Explain how the Start/Finish approach assigns percentages of completion to a task.

2. The Start/Finish approach assigns values to be earned at the start and finish of the task only, based on a percentage of the total value.

3. What are the disadvantages of the Experience/Opinion approach to determining percent complete?

3. The Experience/Opinion approach is highly subjective with very little, if any, tangible or factual information on which to base the percent complete. It is highly subject to human error.

4. The process for determining the progress on an individual task or the project as a whole is called?

4. Earned Value Analysis (EVA).

5. Earned Value Analysis can only be applied to contracts with fixed budgets. True or False?

5. False. It can be applied to variable budget contracts as well.

6. What is the formula for determining the earned value (EV) of a task?

6. Earned Value (EV) = Percent Complete × Maximum Budget (Scheduled Value).

7. Define Quantity Adjusted Budget and explain when it is used.

7. For projects with a variable budget, the quantities of the tasks change as the result of adding or deducting work from the project. As a result, the budget changes to accommodate the changing scope of work. This is called the Quantity Adjusted Budget (QAB).

8. Explain the difference between the ACWP and the BCWP.

8. The ACWP is the dollar amount (or labor-hours) that it costs the contractor to perform the work, while the BCWP is the dollar amount that the contractor earned as a result of performing the same work.

9. What does a Schedule Performance Index (SPI) of less than 1.00 tell the project manager?

9. A Schedule Performance Index (SPI) of less than 1.00 indicates that the work is behind schedule.

10. Is it possible to have a SPI of less than 1.00 and a CPI of more than 1.00? What does it represent?

10. Yes, it is possible to have a SPI of less than 1.00 and a CPI of more than 1.00. It indicates that tasks are being completed behind schedule and it is costing less to do the work than anticipated.

Chapter 8 Analyzing and Reporting Variances in Schedule and Cost

1. The Earned Value S-curve graphs the dollar value of three separate parameters as a function of time. What are those parameters?

1. The BCWS, BCWP, and the ACWP.

2. The Budget at Completion (BAC) never changes from the beginning of the project. True or False?

2. False. For the BAC to remain the same, the actual costs would have to be exactly as planned.

3. The Budget at Completion (BAC) plus the Management Reserve (MR) is called?

3. The Total Allocated Budget or TAB for the project.

4. Explain why the BAC ends at Substantial Completion on the S-curve.

4. The Budget at Completion (BAC) line ends at Substantial Completion because it would not be contractually correct to show that the project ends late before it has started.

5. The completion line on the Earned Value S-curve should move to the right as extensions of time are granted to the contract. True or False?

5. True. The horizontal scale on the graph is time and starts at zero and ends at Substantial Completion.

6. Large initial separations on the EV S-curve between the lines for ACWP and BCWP, with BCWP above the ACWP, may be an indicator of excessive front loading. True or False?

6. True.

7. Earned Value S-curves can be produced for a single task or for the entire project. True or False?

7. True.

8. Explain the purpose of a Control Chart.

8. A control chart tracks past and current performance by regular update and can be used to predict future performance of each indictor with modest accuracy.

9. SPI and CPI charts are used to graph the respective efficiency factors so that they can be tracked over time. True or False?

9. True.

10. Variance thresholds are a measure of tolerance for SPI and CPI efficiencies. Explain.

10. Variance thresholds measure what is an acceptable deviation for both SPI and CPI. It is traditionally agreed upon at the planning stage. Thresholds may vary with the risk that is assessed for various tasks and over the life of the project.

Chapter 9 Recognizing Trends and Forecasting Performance

1. Trend analysis can be used to forecast future labor requirements. True or False?

1. True.

2. There are four considerations that trigger when data becomes an effective basis for forecasting. Define them.

2. The four considerations for using data to forecast are: (1) any learning period for the task must have expired; (2) the crew(s) executing the work must be fixed; (3) the production rate for the task has stabilized; and (4) a minimum of three reporting periods of data has been collected.

3. What are trend charts used for? Give an example.

3. Trend charts are used to track performance and forecast future performance. An example would be three reporting periods indicate Task A is losing money. This data can be used to predict what will happen in reporting period 4 which will be a loss, assuming nothing is changed. It indicates the general direction a task or project is headed.

4. One of the most sought-after forecasted values is the total anticipated cost of the project when it is complete. This is called?

4. Estimate at Completion (EAC). It is an estimate of what the project will cost when it is complete.

5. The amount of money required to complete the project under the current terms and trend is called?

5. Estimate to Complete (ETC). It is the amount of money needed to complete the project from the status date to completion.

6. The Budget at Completion (BAC) less the Estimate at Completion (EAC) is called?

6. Variance at Completion (VAC).

7. A negative VAC tells the project team that the costs are under budget, while a positive value means the costs have exceeded the current target budget. True or False?

   7. False. A negative value for the VAC indicates costs are over budget. A positive value indicates that the costs are under budget.

8. The TCPI is the recommended performance from the status date going forward. For TCPIs greater than 1.00, there is more work remaining than there is budget to pay for it. True or False?

   8. True. It means that the EAC will exceed the BAC.

9. Is an SPI of 1.023 acceptable? Explain.

   9. An SPI of 1.023 indicates that the work is proceeding 2.3 percent faster than planned. It is acceptable by itself. It provides no indication of cost performance.

10. What basic information can be derived from an SPI of 1.02 and a CPI of 0.97?

    10. An SPI of 1.02 indicates that the work is proceeding 2 percent faster than planned. A CPI of 0.97 indicates that the work is costing 3 percent more than planned.

Chapter 10 Productivity

1. Explain the difference between productivity and production.

   1. Production is the measure of output (units of work produced), whereas productivity is the rate of production (units of work per unit of time).

2. Productivity is the measure of output or return for each unit of input. True or False?

   2. True. The input is labor and the output is work.

3. Productivity in construction is uniform and remains constant over the execution of a task. True or False?

   3. False. It will vary daily, but will be represented uniformly over the task as an average for both estimating and scheduling purposes.

4. What is a production model and how is it used?

   4. A production model is a mathematical expression of the production process that is based on collected data, measured in the form of quantities of inputs and outputs. It establishes the basis for measuring performance.

5. The comparison of planned output to actual output is called?

   5. The productivity index or PI. It is another measure of efficiency.

6. Explain what a productivity index (PI) of 1.034 indicates.

   6. A productivity index (PI) of 1.034 indicates that the crew is 3.4 percent more efficient than planned in the estimate.

7. Factors that affect productivity can be separated into external (uncontrollable) and internal (controllable) factors. True or False?

7. True.

8. Statistics show that crews are productive about 95 percent of the workday. True or False?

8. False. The average productivity is far less ranging from 61 to 66 percent.

9. Why are labor-hours a better basis for a productivity index (PI) than dollars?

9. Labor-hours as a standard are not affected by increases in the rate of pay. They remain consistent over time.

10. Identify five controllable and five uncontrollable factors that can affect productivity.

10. Five controllable factors: (1) ambient conditions within an enclosed space; (2) proper planning and task analysis; (3) sufficient materials on hand to perform the work without interruption; (4) repeated interruptions or changes to task scope; and (5) disruptions to work flow by crew changes. Five uncontrollable factors: (1) personal ambivalence or apathy of the workers; (2) outside weather conditions: precipitation, temperature, etc.; (3) personal problems (unrelated to work) diverting attention from work; (4) minor health problems: colds, fatigue, aches and pains; and (5) economic conditions in the region.

Chapter 11 Acceleration and Schedule Compression

1. Define the three types of acceleration.

1. The three types of acceleration are: actual, constructive, and forced acceleration.

2. What is an excusable delay? Provide an example and explain.

2. An excusable delay is beyond the control of the contractor. Inclement weather that prevents the work from proceeding is an excusable delay.

3. Explain why most contracts provide that a general contractor can supplement a subcontractor's workforce even if the sub is an independent entity.

3. Contracts provide that the contractor has the responsibility and the right to fulfill their contractual obligations even if a portion of the work is subcontracted. In addition, most contracts between the owner and the GC do not contractually recognize subcontractors.

4. Define the term "Schedule Compression."

4. Schedule compression is the shortening of the duration of a task or the shortening of the entire project.

5. Schedule compression rarely results in the increase of labor resources or labor-hours. True or False?

5. False.

6. What is time-cost trade-off analysis?

6. The process by which the project team decides which tasks offer the most time savings for the least cost.

7. Explain the difference between a direct cost and an indirect cost on a project.

7. A direct cost is directly related to one project and that project only. An indirect cost is a cost relative to operation of the company and applies to all projects.

8. Explain the term "crash duration."

8. Maximum schedule compression is called crashing. As a result, a crash duration is the shortest amount of time in which a task can be performed.

9. It is the purpose of the time-cost model to identify the task with the least cost per crash day. True or False?

9. True. It results in a cost per day saved.

10. Why are contractors reluctant to produce a recovery schedule?

10. By producing a recovery schedule a contractor is admitting to being behind schedule and thereby opening themselves up to a claim for delay.

Chapter 12 Resource Management

1. What are the two goals of resource management?

1. Resource management has two key goals: (1) to ensure that the correct resource is available at the correct time and in sufficient quantity; and (2) to ensure that the resource is used efficiently.

2. Labor resources can be divided into two types. What are these two types?

2. Hourly and salaried wages.

3. Material resources can be divided into two types. What are these two types?

3. Materials can be: support materials that are not incorporated in the work, but provide support for the work, or installed materials that are permanently incorporated in the work.

4. Scheduling and coordinating resources so that they arrive on site as needed and removed when not needed is called?

4. Resource allocation.

5. It is the goal of resource allocation to maintain sufficient resources to allow the work to proceed in a smooth or even flow. True or False?

5. True.

6. Define the term "time-constrained project."

6. A time-constrained project assumes that the resources available to do the project are sufficient, or can be supplemented to complete the project within the time available. Time is the constraint.

7. It is not the goal of the resource leveling process to make the daily utilization of labor, as uniform as possible. True or False?

7. False.

8. What is one disadvantage of resource leveling in construction?

8. The disadvantage of resource leveling is the redirected focus from schedule delivery to resource absorption, thereby overriding the critical path.

9. What can the project manager expect to see from a resource profile?

9. A resource profile shows graphically the day-by-day demand for a particular resource based on a defined schedule for the project.

10. The actual management of materials can be separated into two competing schools of thought. Explain them.

10. The actual management of materials: (1) deliver the materials just prior to installation; or (2) provide an adequate supply of materials to be maintained on site.

Chapter 13 Risk Management

1. All risk can be eliminated from a project through aggressive mitigation plans. True or False?

1. False. It is impossible to remove all risk.

2. Risk can have a positive or negative effect on the project goals. True or False?

2. True. We exploit positive risks if possible and minimize negative risks.

3. What is the purpose of the risk management process?

3. The purpose of the risk management process to identify potential risks, reduce their chance of occurring, and minimize their impact if they do occur.

4. The first step in the risk management process is risk identification. True or False?

4. True.

5. Probability is generally considered a more detrimental factor than impact. True or False?

5. False. The impact caused by the event is more detrimental.

6. **What is the purpose of an impact scale?**
   6. An impact scale provides a numerical rating for judging the severity of the risk against predetermined factors.
7. The impact scale can then be used as the basis for a comparison that allows the project manager to assemble and rate the risk based on a numerical scale. This scale is called?
   7. The risk severity scale.
8. The Failure Mode and Effects Analysis (FMEA) introduces what additional factor to the severity matrix?
   8. FMEA introduces detection to the severity matrix.
9. The process to reduce the likelihood that the event will occur and the impact of the event if it does is called?
   9. Risk mitigation.
10. PERT analysis of a task is used to identify the most probable duration of the task. True or False?
    10. True. It uses probability to determine the best duration.

Chapter 14 Project Closeout

1. What is the objective of the project closeout process?
   1. The objective of the closeout process is to create an orderly and documented transition of the project, from possession by the construction team to possession by the owner or end-user.
2. Project closeout begins at the end of the project. True or False?
   2. False. It begins in the pre-planning stages of the project.
3. Project closeout procedures can be separated into two main categories. Define each.
   3. Project closeout procedures can be separated into two main categories: (1) contract closeout and (2) administrative closeout. Contract closeout is the reconciliation and review of the contract to determine if everything has been completed. The administrative closeout deals the collection and archiving of administrative documents.
4. Lessons learned meetings are used to avoid repeating the same mistakes and to improve performance on the project. True or False?
   4. True. Lessons learned meeting are intended to be a review of what was done well and what was done poorly over the life of the project.
5. Explain how job cost reports are part of the project closeout and lessons learned meeting.
   5. Job costs reports represent the cost component of the administrative closeout.

6. Name five items that one would expect to see in lessons learned documentation.

6. (1) Describe in sufficient detail the issue that occurred; (2) summarize any triggers or thresholds for recognizing the issue was occurring; (3) outline in detail the impact (cost and/or schedule) of the issue; (4) outline in detail the root cause of the issue; and (5) provide references or sources of information for diagnosing the issue or applying the corrective actions (if a negative issue).

7. Why is it essential for the report to contain viewpoints other than those of the project manager?

7. So that the report is evenly balanced and less biased by opinions of the person who managed the project.

8. What is the purpose of archiving files?

8. To store them for future reference.

9. Explain where a project team member would find what is required for closeout.

9. Refer to Section 01 70 00–Execution and Closeout Requirements in Division 1–General Requirements section of the technical specifications.

10. The estimator(s) should be part of the project closeout process. True or False?

10. True. There is historical data that can be of benefit to the estimator for next project of a similar type. It can also be a time to comment on the estimate itself.

Index

Project Control: Integrating Cost and Schedule in Construction, Second Edition. Wayne J. Del Pico.